Continuous Improvement
in the
Mathematics Classroom

Also Available from ASQ Quality Press

Improving Student Learning: Applying Deming's Quality Principles in Classrooms
Lee Jenkins

Continuous Improvement in the Science Classroom
Jeffrey J. Burgard

Continuous Improvement in the Primary Classroom: Language Arts K-3
Karen R. Fauss

Thinking Tools for Kids: An Activity Book for Classroom Learning
Barbara A. Cleary, Ph.D. and Sally J. Duncan

Futuring Tools for Strategic Quality Planning in Education
William F. Alexander and Richard W. Serfass

Quality Team Learning for Schools: A Principal's Perspective
James E. Abbott

The New Philosophy for K-12 Education: A Deming Framework for Transforming America's Schools
James F. Leonard

Creating Dynamic Teaching Teams in Schools
K. Mark Kevesdy and Tracy A. Burich, with contributions from Kelly A. Spier

Success Through Quality: Support Guide for the Journey to Continuous Improvement
Timothy J. Clark

To request a complimentary catalog of ASQ Quality Press publications, call 800-248-1946, or visit our online bookstore at qualitypress.asq.org .

Continuous Improvement in the Mathematics Classroom

Grades K–6

Carolyn Ayres

ASQ Quality Press
Milwaukee, Wisconsin

Continuous Improvement in the Mathematics Classroom
Carolyn Ayres

Library of Congress Cataloging-in-Publication Data
Ayres, Carolyn, 1943–
   Continuous improvement in the mathematics classroom. Grades K–6 / by Carolyn
Ayres.
      p. cm.
   Includes bibliographical references and index.
   ISBN 0-87389-432-4 (alk. paper)
   1. Mathematics—Study and teaching (Elementary) I. Title.

QA135.5 .M465 2000
372.7—dc21   99-057218

10 9 8 7 6 5 4 3 2

ISBN 0-87389-432-4

Acquisitions Editor: Ken Zielske
Project Editor: Annemieke Koudstaal
Production Administrator: Shawn Dohogne
Special Marketing Representative: David Luth

ASQ Mission: The American Society for Quality advances individual and organizational
performance excellence worldwide by providing opportunities for learning, quality improvement,
and knowledge exchange.

Attention: Bookstores, Wholesalers, Schools and Corporations: ASQ Quality Press books, videotapes,
audiotapes, and software are available at quantity discounts with bulk purchases for business,
educational, or instructional use. For information, please contact ASQ Quality Press at 800-248-1946,
or write to ASQ Quality Press, P.O. Box 3005, Milwaukee, WI 53201-3005.

To place orders or to request a free copy of the ASQ Quality Press Publications Catalog,
including ASQ membership information, call 800-248-1946. Visit our web site at www.asq.org or
http://qualitypress.asq.org .

Printed in the United States of America

 Printed on acid-free paper

American Society for Quality

Quality Press
600 N. Plankinton Avenue
Milwaukee, Wisconsin 53203
Call toll free 800-248-1946
Fax 414-272-1734
www.asq.org
http://qualitypress.asq.org
http://standardsgroup.asq.org
E-mail: authors@asq.org

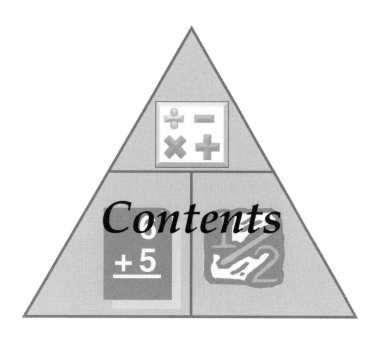

Contents

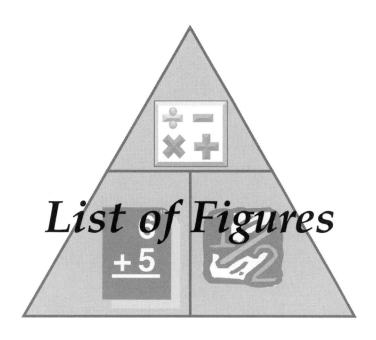

List of Figures

Chapter 2

Chapter 3

Chapter 4

Chapter 5

Chapter 6

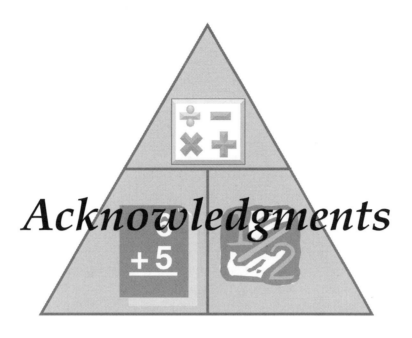

Acknowledgments

Dr. Lee Jenkins is responsible for my immersion into mathematics. Before I believed in myself, he believed in me. He encouraged me to become a math mentor and a presenter at a northern California math conference. With his guidance, I began to see myself as a teacher of mathematics.

Dr. Deming said, "Learn from the masters; they are few." Dr. Jenkins has been one of those few. So many times, my view of how to do things, or how to think about something has changed after visiting with him or attending one of his workshops. For example, on one occasion, I spent a number of hours preparing an after-school math presentation for teachers in the district. Dr. Jenkins was one of only four people who attended. I felt discouraged, thinking of all the time I had spent preparing, only to have these few participants. My view of how to think was changed when Dr. Jenkins told me that even if I helped just one person to improve something he/she was doing, I could consider my time well spent. This one person would influence another and that one another; change would come about as a result of my efforts. In the years following, whenever I saw a small group of teachers in front of me, I relaxed and looked forward to the fun. I remembered I was making a difference in the lives of individuals. I have repeated these thoughts to other presenters who have needed a fresh look at what it means to have just a few attend a workshop.

Dr. Jenkins sees the potential in others and empowers them to rise to that potential. When he asked me to write this book, he saw in me a potential author. I thank him for the opportunity and adventure this book has been.

xiv   *Continuous Improvement in the Mathematics Classroom*

My daughter, Shawni, is another I have to thank. She is a high school English teacher with a brilliant and inquiring mind. As she helped me edit she dispensed with warm fuzzies, and went right to routing out fluff, confusing language, and poor reasoning. I honor her honesty. I thank my son, Dave, who is on a journey of interpersonal growth. Because who I am cannot be separated from my teaching, my long visits with Dave about manifesting love and abundance into our lives, has enhanced the peace and joy I feel as a teacher. I honor Dave for who he is and the inspiration he is to me.

I thank Tom Fox, English professor at Chico State University, and director of the Northern California Writing Project. Under his guidance, I evolved from being a teacher of writing to being a writer who teaches. As I began writing this book, he was able to help me see I wanted to write a personal account of my teaching, not a research paper on Dr. Deming.

My thanks to Bill Fisher, director of the Center for Mathematics and Science Education at Chico State University, and overseer of the CSU Chico Mathematics Project. Through my participation in the Mathematics Project, I gained confidence in my own ability to think.

Ron Johnson, a sixth-grade teacher, and Pam Stephenson, a fifth-grade teacher, each willingly shared teaching and classroom management ideas with me. Each is a gifted educator with a love for teaching.

Judy Flores, a district math consultant, was responsible for putting together many of the problem solving and math assessments I used in the classroom. When I ran into mathematical snags, she willingly shared her expertise. I thank Peggy McLean, author of many books, as well as presenter and math consultant, for the forty-five minute period she spent in my classroom right at a time when I needed new eyes. Her suggestions began a change of direction in my teaching.

I thank my students for their enthusiastic participation in the math activities of which this book is about. Because I got up each morning between 3–4 A.M., I often whined that if I didn't have to teach, I would have more time to write. Following that thought came: "But if I didn't teach, I wouldn't have anything to write about."

My last acknowledgment is to Gary Mercer, my husband, who was a sounding board for my good and not so good ideas; who slept alone half of every night while I sat pecking on the keyboard; who helped me with my computer problems; and who made me laugh at least once each day.

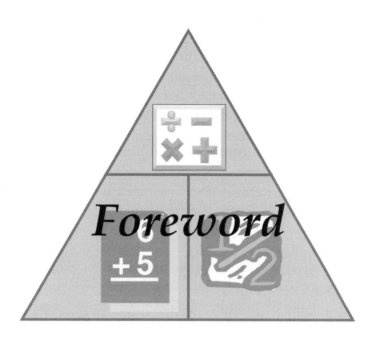

Foreword

by Lee Jenkins

The books in the ASQ Continuous Improvement Series are written for teachers who possess both the utmost respect for their students and a desire to improve student learning. Teachers are often frustrated because the tools society provides for student improvement do not honor their respect for students.

Four currently popular methods for improving student learning are these: (1) add more fear to the lives of students; (2) bribe students with incentives to do better or more work; (3) set up false competition between students; and (4) purchase a new program and demand its use.

There are, however, thousands of teachers who love and respect their students and recognize that fear, bribery, false competition, and my-way-or-the-highway approaches don't work. At best they achieve short-term results. The books in the ASQ Continuous Improvement Series, by Fauss, Burgard, Carson, and Ayres, are written from the hearts of teachers who possess the two criteria mentioned in the first paragraph. They love and respect their students while maintaining an intense desire to improve their students' learning.

These teachers clearly describe their experiences with a fifth option: studying and using data for continual improvements. They document the systems their classrooms use to collect weekly data, as well as the process of making curricular decisions based upon their data. The data is from students' long-term memories. There is no place for cramming and short-term memory. The teachers and their students, functioning as a team, can and do plan their improvements. They know within a few weeks if the instructional plans are working or not. No longer must these teachers wait until July to see results.

Readers will enjoy these teachers' experiences and stories, but most of all they will learn exactly how to restructure the management of their classroom learning systems toward significant improvement.

Each of the four authors in the ASQ Continuous Improvement Series brings a unique view of quality processes into his or her classroom.

Fauss: In the culture of education, teachers call themselves lucky when they have an exceptionally bright class. The unstated inference is that the next year will return to normal. What happens, however, when a teacher is determined to prove that every year will end with better results than the previous year? After a year or two, this becomes intense, because the easy ideas are used up. Fauss incorporates dynamic language arts methods with the tools for continuous improvement. Talent and passion will transfer from her pen to the minds and hearts of readers.

Burgard: It is clear that Burgard sees the application of quality principles through the eyes of a scientist. He changed the English department's rubric into a dichotomous key, and he analyzed student writing errors like a chemist. Further, he knows science is not science unless it can be replicated. This means students must have precise knowledge of definitions and clear understanding of scientific processes. Both information and knowledge are essential; there's no room for an educational pendulum in Burgard's thinking.

Carson: Most surveys of school attitude document that history/social science is the most hated of school subjects. Readers of Carson's book will easily see, however, why history is her students' favorite subject. Not only are they loving to learn the facts of history, they are thinking like historians. She pushes student involvement in planning further than I've ever seen—even including scheduling lessons and assignments. This is not just the Carson show—it's better.

Ayres: This book combines the principles of quality with research taken from the Third International Math and Science Study (TIMSS). As a part of this study, videos were made that demonstrate Japanese methods of teaching mathematics. Ayres has combined these Japanese methods with principles of quality. Her work is truly unique in the world of mathematics education. Further, she documents her students' growth in mathematics concepts weekly, and their attitude toward mathematics monthly. Her formula for success is this:

> *mathematics concepts + mathematics problem solving + great attitudes + the data to prove each is occurring = great school mathematics*

The poorer an idea, the more age specific it is; the better the idea, the wider the age span it has. Even though these four teachers each teach a particular grade and relate stories about a particular age group, audiences from kindergarten through college may benefit from their provocative lessons.

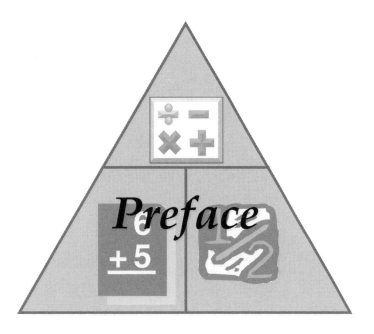

Preface

In this book, the reader peeks through my eyes to watch some new concepts unfold in the classroom. The application of ideas covered go beyond the scope of the grade level that I have been teaching while writing. Although the math activities are most appropriate for grades K–6, many ideas probed are timely for teachers of kindergarten through twelfth grade, and for subjects other than mathematics. My attitudes about what math teachers could learn from each other were broadened when I attended the CSU Chico Mathematics Project, an intense two year in-service training that included K–12 teachers. I sat beside high school teachers as we problem solved and talked together, shared lessons and resources. Many math activities I came into contact with were easy to revise and recreate to fit my own teaching situation. I invite the reader to do the same with the ideas presented in this book.

I teach in Redding, California, at Alta Mesa School in the Enterprise School District, a K–8 school of about 600 students. I did my student teaching in a junior high and started my teaching career at Klamath Falls High School. In my twenty-four years of teaching, I have taught in every grade K–12 in one capacity or another. I was teaching second grade at the time of this writing. It is a class I *looped* with, which means I started with the children in first grade, and then moved with them to second. Teaching at many different grade levels has helped me realize there is no chosen grade level, one that gives status to the teacher teaching it. Every teacher plays an absolutely essential role in creating the end product—competent high school graduates.

Mathematics is not the only subject I love to teach. I have given many teacher workshops integrating the teaching of language with rhythm and song—much, much fun! Teaching is not about a set of things to do with a given age of children. It is about interacting with the goals, the appropriate information and knowledge, and the students, in a way that brings excitement and growth; it's finding where the students are now, so they can be moved from there to the next rung on the ladder of learning.

It is hard for me to choose which is my favorite chapter in this book. If I had not done the thinking and experimenting that brought about the first one, "Tracking Student Growth with Data," this book would not have been written. I begin by using some learning activities I have never taught before. The reader is taken through my trial and error process as I use quality measurement, chart the input, evaluate, change, and begin again. I communicate the students' enthusiasm as they work together toward a common goal; as they realize how much fun mathematics can be; as they see they are improving. The first chapter, "Tracking Student Growth with Data" is mostly about the area of learning called knowledge, which translates as problem solving. The second chapter, "Deciding What to Teach," emphasizes the information area of learning—facts, computation, vocabulary. In this chapter, the reader is presented with a range of resources for making decisions about what to teach. Chapter 3, "Managing the Classroom," gives ideas for organizing math manipulatives, working with small and large groups, collecting and checking papers, monitoring behavior, meeting the needs of students with high and low ability, and using math centers to provide hands-on practice. Chapter 4, "Forming a Relationship with the Most Important Supplier," provides some ideas for forming a successful partnership with parents. I show that parents can be counted on to be driving forces in their children's learning. In chapter 5, "Making Connections," we look at relationships between math strands. We also see examples of integrating math with other subjects, like P.E. and social studies. Hopefully, it is just a beginning in sparking other connections that can be made. "Reflecting" is the final chapter. I look back on my school year to see what worked well and what should be changed.

May the reader have as much fun reading this book as I have had writing it.

Introduction: Laying New Track

Whatever you can do, or dream you can, begin it.
Boldness has genius, power, and magic in it.

GOETHE

In my various life experiences, a truth has emerged: If there is something I want to do or have, I can do or have it if I can clearly define it. If I can create it now in my experience as I wish it to be in the future, I become motivated with its impending reality. I find myself culling out things that do not matter. I select things that enhance my journey. What I do on the outside is aligned with that which I believe is possible from the inside. This creates an integrity—my outside working with my inside. Integrity feels good. I'm excited, alive! My created yearning, along with what I will call cosmic intelligence guides me. When that which I envisioned emerges into actuality, I give thanks, celebrate my creation, and begin again. What do I want to do now?

Psycho-Cybernetics, a book by Maxwell Maltz[1] that I read back in the 1970s was my introduction to pulling the future into the present. It had been so many years since I read the book, I wanted to revisit the passage that made such an impact on me. I found the book, and sure enough, there it was—the description of a self-guided torpedo, or interceptor missile—just as I remembered, and still as timely. Maltz's explanation of what a self-guided torpedo does, is a picture of me, angling my way toward a target; and as you will see, also a picture of my students aiming, adjusting, aiming, adjusting toward their targets. Here is the passage I found:

> "The target or goal is known—an enemy ship or plane. The objective is to reach it. Such machines must 'know' the target they are shooting for. They must have some sort of propulsion system which propels them forward in the general direction of the target. They must be equipped with 'sense organs' (radar, sonar, heat perceptors, etc.) which bring information from the target. These 'sense organs' keep the machine informed when it is on the correct course (positive feedback) and when it commits an error and gets off course (negative feedback). . . . The torpedo accomplishes its goal by *going forward, making errors,* and continually correcting them. By a series of zigzags it literally 'gropes' its way to the goal.
>
> "Dr. Norbert Wiener, who pioneered in the development of goal-seeking mechanisms in World War II, believes that something very similar to the foregoing happens in the human nervous system whenever you perform any purposeful activity . . ."

At the beginning of the school year, it was time to ask, "What is my target? Where do I want to go?" The more clearly I could define my target, the more likely it was I could get there. Many things came to mind. It is easy to get lost in all of the learning objectives for the subjects that must be taught. Intuitively I have always known that to teach anything, students must first be motivated. Bringing this out into the conscious realm, I chose to work again with a well-used theme: Create a classroom in which the students are motivated to learn; one in which they take responsibility for their own growth.

As I was planning to do this, a gift came into my hands—a book: *Improving Student Learning: Applying Deming's Quality Principles in Classrooms* by Dr. Jenkins.[2] I had a particular interest in reading the book because Dr. Jenkins was the superintendent of the district in which I had recently been hired to teach. From Jenkins' book, I learned that Dr. Edward Deming was an American physicist, mathematician, and statistician, who began going to Japan in the 1950s to work with businesses. His philosophies have been responsible for helping those businesses to become high-quality, low-cost enterprises. Since that time, many American businesses have benefited from what is called Deming's Philosophy of Profound Knowledge. Dr. Jenkins' book shows how this wisdom can be used in a school district.

What so fully captured my attention was the motivation built into Dr. Deming's system—any system, business or education. He showed that when a clear aim is established, and data is used to track the movement to the aim, those whose aim it is become excited and motivated; and they work together because they realize they get there faster by helping each other. Everyone takes responsibility for the quality of what is being produced, and quality is the one thing that is never sacrificed. Dr. Deming challenges the misconception that quality costs more. He shows that, given enough time for a system to become stable, quality will always cost less. Thousands of dollars are wasted in the California educational system, but in my classroom, the biggest cost I am faced with is time. I race the clock each day. I race the calendar each week and month. Will the children know what they need to know by the time I see them out the door on the last day of school? I am continually aware that if I can improve the quality of my teaching, I can save valuable time.

I got excited about Edward Deming's ideas as presented by Dr. Jenkins so I decided to try some of them. What has come about is a fascinating alignment between Maltz's psycho-cybernetics and Dr. Deming's ideas for data collection. Students are using data to monitor their growth as they move toward a target. They are motivated to learn as they watch a line graph of class scores climb two steps up and one step back, wiggling its way toward the top. The rudder Deming uses is called "Plan, Do, Study, Act." In the classroom it looks like this: After a clear aim has been established, the teacher, along with the students, chooses an area of growth, something to improve. They make a plan on how to do this. The teacher guides the planning, basing choices upon her best knowledge of how students learn. This is the first step. For the second step, students begin doing, carrying out the change on a small scale so as to maintain the present stable learning conditions. The small scale is important. In business, when huge changes are made every time the data indicates a movement away from the aim, Dr. Deming calls it tampering. Tampering is jumping to change things every time the data show a drop away from the goal, rather than allowing the usual amount of movement required to let what is being changed become stable. In the classroom, huge pendulum swings are disruptive to learning.

For the third part of Deming's formula (Plan, Do, Study, Act), data are collected and studied. From this feedback, the teacher and students, brainstorm ways to improve. Changes are made and the cycle begins again. Using this method to stay continually aligned with the aim is the kind of quality teaching that could cost less time. Dr. Deming said quality is not about working harder; it's about working smarter. It is about connecting to the aim, and producing the quality results that align with that aim.

Think of the time saved when we as a classroom of people, know where we're going and we're helping each other to get there. When students see the class's growth as shown by data, and they realize they have something to do with having the line on the chart go up or down, or whatever way it is going to show improvement, it turns into a game, and students love games. They are motivated. Individual growth is charted and shared personally. Class charts are made visible for motivation and celebration. As students work to improve their own skills and knowledge, their efforts benefit the whole class. When the line drops on the graph, the class, along with the teacher, looks for causes. Solutions are brainstormed and changes are made. The feedback guides the instruction; and the feedback motivates students to practice, to do that which is needed to bring about improvement. All are involved in the process of moving the line upward.

In the past, after a week of doing learning activities in my classroom, I have wondered, "Were we going anywhere? We were content. We had a good time, but one more week of the school year is gone. Was that week pointed in the direction I wanted to be going?"

This reminds me of a story. Many passengers are on a train, happily chatting and passing the time, unaware that the train is taking them toward disaster. Someone looks out the window, and exclaims, "We need to get out and lay some new track or we will end up where we're headed!"

Interacting with current data regularly, not only has me looking out the window and laying new track, it has the students eagerly helping. Children tend to be less goal oriented than adults, unless getting in enough play time can be seen as a goal. Without the line graphs, students are perfectly willing to let me figure out what is supposed to be done. For them, the future is a fuzz. When they get involved with the data, they are motivated to help lay the new track.

Let me carry this metaphor a little farther and suggest that the cars on the train are the schools or classrooms in a district. I cannot have the success I desire if my aim is different than those of other teachers at my school, or of the district as a whole. The aims of the classroom must nest inside the aim of the whole system. We must all look carefully to see where we want to go, and then work together to get there. With the guidance of Dr. Jenkins, the Enterprise School District chose an aim that encapsulates Deming's philosophy: "Maintain enthusiasm while increasing learning." What I am doing supports and defines that aim.

The chapters of this book can also be seen as cars on a train. As conductor I must look carefully at my destination. Will I be pleased when I end up where I am headed? As I asked this question, there came several occasions in which I had to get out and lay new track. The problem at first was my own confusion about my aim. The working harder not smarter parts sit in the trash can. With continued commitment to defining my direction, my purpose for writing this book crystallized: I want to create an impelling picture of the brilliance of creating student enthusiasm for learning through the use of data to track growth. As a part of this, I want to talk about how to get the growth— what can be done to guide and facilitate the movement up the chart, the groping to the target. I have collected some of the best ideas I know for teaching mathematics, ones I have tried myself, and ones that other teachers have graciously shared with me. With all of this, I hope the reader will be motivated to lay some new track of his/her own.

This mental grunt work required to establish an aim and then find ways to align with that aim is worth the effort. A focused core of thought allows ideas to spring forth that are more likely to support the intent, pure and clear. If I want an apple tree, I plant an apple seed. Too many times I have planted a seed that grew *something,* without wondering if it was what I really wanted. The mental grunt work goes into determining the kind of seed to be planted. Then all the work that follows becomes rewarded because the end product is a masterpiece of my desire.

When my husband and I bought three flat, treeless acres to begin a new home, we began planting trees even before the house was finished. I never questioned Gary's desire to put in rows and rows of walnut trees. For what seemed like months, I went out and helped dig holes. Three years later I laughed out loud when Gary turned to me and said, "Remind me. Why was it we planted all these walnut trees?"

After reading Dr. Jenkins' book, I wanted more of the story, so I invested in some of the other books written on Dr. Deming and his philosophy for creating a working, successful system. If the reader's interest becomes similarly piqued, some of these books are listed in the references. Dr. Deming's Fourteen Points can be found in the appendix. My own book, besides being one that provides practical ideas for teaching math, is about my interaction with the thoughts of Dr. Deming. A summary of those thoughts will provide a foundation of understanding for my own thinking.

To understand Deming's philosophy, as applied to education, it is important to know what is meant by *system.* Although they may manifest differently, the same parts exist for all systems. The seven components are aim, customers, suppliers, input, process, output, and quality measurement. In his book, Dr. Jenkins translates what each of these mean in terms of a school system. Here is my own capsulated version: Children can be thought of as *customers,* the ones who receive the intended product—knowledge. Ultimately, the customer must be pleased. A child's whole educational environment can be thought of as the *input* that feeds the child with the learning necessary to mold him/her into the desired *output,* which is a well-rounded, happy, contributing member of society. *Supply* is education the child has received in prior years, from parents, the community, and past teachers. The *aim* is the umbrella under which all the parts exist. It pinpoints exactly what the system is all about. The aim is created and embraced by all those who work under its unifying effect. *Input, process, output,* and *quality measurement* are the gears that move the system forward in its purpose, which is to create continual improvement (see Figure 0.1). Input for improvement comes from all possible sources: management, teachers, children, classroom aides, bus drivers,

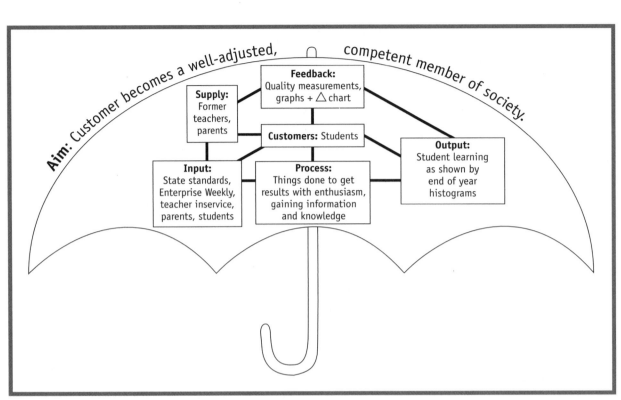

Figure 0.1.   The Educational system.

board members, parents, cooks, custodians, and members in the immediate, and far-reaching community. Also, the system cannot survive without the input that comes from quality measurement. Using input, processes are tried and revised continually. Feedback comes from the output. Changes result in the improvement of the product and the service. Rather than the traditional organizational pattern where decisions are made and ensuing activities are dictated by management, all those who work in the system are respected and valued for their contributions to the good of the whole. Management does not evaluate, and compare one member to another. Instead, programs of self-improvement and education build responsibility and leadership. Collaboration rather than competition creates all winners and no losers. Joy, not fear, permeates the workplace. Rather than resorting to quick-fix methods, workers see that long-range goals are what hold the promise of lasting improvement. Quality is integral to everything that is done, because all participants have the knowledge that, besides the joy that comes from a job well done, commitment to quality saves time and money.

Once a better way of doing things becomes evident, there is no going back to Kansas. That's the way I felt after becoming aware of the utopian school system that could be built based upon Deming's principles. The body of knowledge necessary for such a system is profound, but simple—like the wheel. Once the concept of the wheel came into being, I'm sure people wondered why it had not been thought of sooner, and to continue life without it made no sense at all. Once Deming principles are understood, it makes no sense to set up a system any other way.

A group shift in thinking is required for such a system to come into reality. If that shift doesn't take place, carrying out part of the ideas will not bring the growth and success desired. Granted, all the ideas are good ones, but without the shift, they are implemented in isolation. Once I was given a paper that said "Read all the directions before you begin." I then carried out everything I was told to do, numbers two through ten. Number eleven said, "Follow only the directions to number one." In Deming's system, number one says, "Do everything, or go back to things as usual." If, after being advised of this, one sets up just part of the system, the effort is pretty much wasted. Imagine someone trying to use a wheel that has a long flat side, and then complaining that it doesn't work. That person would have something in common with the one who uses just some of Deming's good ideas, and then complains that a Deming system doesn't work.

With these ideas in mind, let's get on the train and go for a ride.

Notes

1. Maxwell Maltz, *Psycho-Cybernetics* (Prentice-Hall, Inc.: Upper Saddle River, NJ, 1958), 19–20.
2. Lee Jenkins, *Improving Student Learning: Applying Deming's Quality Principles in Classrooms* (ASQ Quality Press: Milwaukee, WI, 1997).

Figure 0.2.   Dr. Deming. Drawn by Tanner Mercer, age 5.

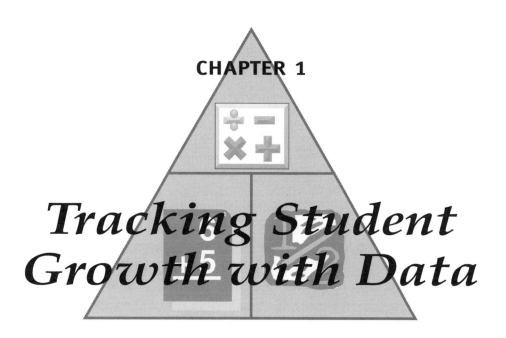

CHAPTER 1

Tracking Student Growth with Data

I lay out the math smorgasbord and the children either eat heartily, or they pick at their food. When I laboriously poke concepts down the students' throats, growth is slower than if they eat because they are hungry. My greatest calling as a teacher is to create the hunger. I intend to show that including students in the data collecting process is a motivational engine.

First, a few thoughts on motivation itself. In his later years of lecturing, Dr. Deming spoke of the educational system. He said all children are born motivated to learn. Our job as teachers is to create situations that will have them keep their yearning for learning. What?! They are born motivated to learn? Why am I putting so much energy into finding ways to motivate then? I've always thought motivating was part of my job. I have to admit that when I observe young children, I see their strong motivation to learn to walk and talk, to jump and climb. But thinking of them as coming to school, still in this state of wanting to learn, causes me to reorient my thinking. I have observed that each year *some* students show up motivated to learn. But Dr. Deming said *all* children are motivated to learn. If this is really true, it means that I can look around my classroom, find those who seem to be lacking their drive, and stop doing those things that demotivate them. What a thing to ponder. How do I know when I am demotivating my students? David Elkind wrote, "One of the most serious and pernicious misunderstandings about young children is that they are most like adults in their thinking and least like us in their feelings. In fact, just the reverse is true, and children are most like us in their feelings and

1

least like us in their thinking."[1] Motivation is based in feeling. So if I remember that whatever demotivates me demotivates the child, I can ask myself, How would I feel if the principal used this technique with me? If what I'm doing with the children would demotivate me, I have a clear message. Stop. It is the Golden Rule reapplied. "Do unto the children as you would have done unto you."

Ranking

This brings me to the subject of grading and report cards. I read in Dr. Jenkins'[2] book, "Ranking destroys joy. . . . The major way educators rank is with grades. Grades destroy joy. Destruction of joy destroys learning." In an e-mail communication in which I had asked some questions, Dr. Jenkins wrote, "Ranking is one of the instruments that destroys intrinsic motivation. Anything which causes a student to think he is less than capable, which ranking does, makes the student feel less whole." I think a parent's disappointment in his/her child's ranking can be an added cause for demotivation of a student. Like many teachers, I have never liked giving grades, but have had no viable alternative. Parents want to know how their children are doing. How do I show this without using grades?

A number of years ago, the Enterprise School District began to create its own measurements to show growth in the areas of reading, spelling, writing, and math. By the time I re-entered the district, after having been gone for eight years, the tools had been perfected and were being used district-wide. The desire was to create a working educational system based on Deming principles. One of the parts of a Deming system is quality measurement. Enterprise District's intent was to develop some tools for quality measurement. Mid-year and end-of-the year tests were given at each grade level, and these numbers were measured against standards that had been established. After the final tests, feedback was given to parents and teachers in the form of graphs (see Figure 1.1). This was the evolutionary work that brought the first and second grade teachers at my school to a place where we were able to create report cards that do not give grades, but instead compare students' growth to district standards (see Figure 1.2).

Up to this point, I had not thought through why using grades ranked students, and using benchmarks or standards did not. Researching what is exactly meant by a letter grade gave me my answer: The theory behind letter grades is a forced curve. Only a few can have As, and a few must have Fs. If, instead of a curve, there is a standard for an A and everyone, or almost everyone, earns an A, then grades are the same as standards. In all of my years of teaching, I have followed the mass consciousness, and graded on a curve, not knowing any alternative.

Enterprise School District—Second Grade Assessments

ASSESSMENT SCORE FOR: *KAREN SMITH*

Math Concepts

In this test, students solve a variety of math problems. These problems assess students' abilities to compute, measure, and show ability with other math understandings. There are 40 items on this second grade test. In order to meet Standard (grade level), students need to have 33–36 problems correct.

Karen correctly answered 24 of the 40 problems.

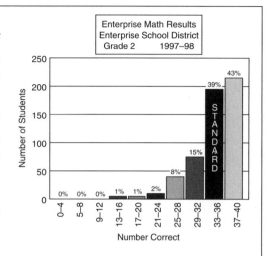

Writing

In this assessment, students respond in writing to a prompt shared by the teacher. The score is based on a six-point scale, six being the highest score. In order to meet Standard (grade level), students need to have a score between 3.5 and 4.5.

Karen's score on the writing assessment is 2.5 on a scale of 1 to 6.

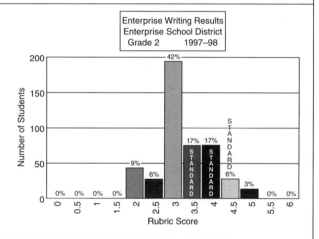

Spelling

Students, by the end of 8th grade, are asked to spell the 1000 most frequently used words in English. At second grade, students are given a random set of 40 out of the first 200 spelling words. In order to meet Standard (grade level), students need to have 30–36 words correctly spelled.

Karen spelled 24 of 40 words correctly.

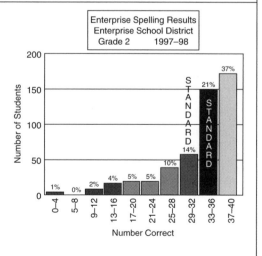

Figure 1.1. Enterprise school district—second grade assessments. Used with permission of the Enterprise School District.

1998–99 REPORT CARD—ALTA MESA SCHOOL

Second Grade

Name: _____

Teacher: _____

	Fall	Mid-term	End-of-Year
READING			
Book level/accuracy	_____	_____	_____
Standard 90%	Book 7 (70wpm)	Book 8 (75wpm)	Book 9 (90wpm)
High frequency words	_____	_____	_____
Standard	120–150	150–175	200
SPELLING			
Weekly tests			
Standard 90%	_____	_____	_____
Grade level:			
Bear tests	_____	_____	_____
Standard	3	6	10
200 word list	_____	_____	_____
Standard	50%	70%	90%
WRITING			
Rubric score	_____	_____	_____
Standard	3.0	3.25	3.5
MATH			
Weekly scores	_____	_____	_____
Standard	50%	65%	80%
Problem solving	_____	_____	_____
Standard	40%	55%	70%

Figure 1.2.   1998–99 second grade report card—Alta Mesa School.

Besides taking away the aspect of ranking, the information I was able to give parents on their child's new report card was more telling than any I had been able to communicate in the past when using the traditional report card. Why? Exact information and percentages were given based on the quality measurements. Parents could see where their children were in relationship to each subject's standard. Rather than a ranking, it was, How far

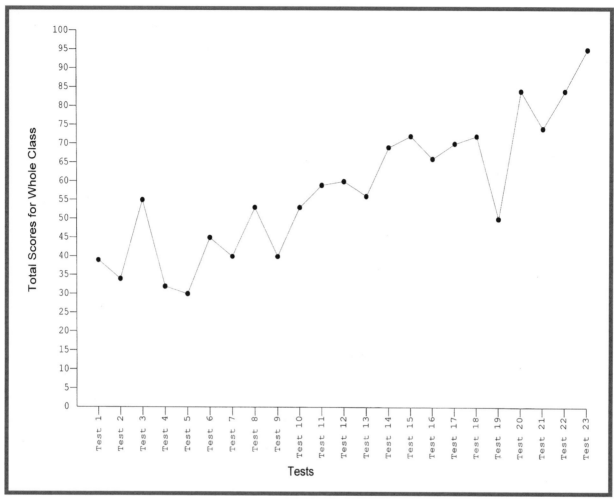

Figure 1.3. Enterprise Weekly tests—class run chart.

is my child from meeting the benchmark? A computer program that was created for tracking student growth gave me a clear way of presenting the results of these quality measurements.

This computer program, called Class Action[3], is not only an effective way to communicate with parents; it communicates information to students, and it communicates information to me. If I were the only person to ever see the graphs produced when I enter the numbers collected from the assessments, the program would be worth all the thinking and planning that has gone into creating it.

In the accompanying examples you will see a number of ways information can be displayed using this program: Class Run Chart (see Figure. 1.3), Student Run Chart (see Figure 1.4), and whole class Scatter Matrix (see Figure 1.5). At the end of this chapter I show a histogram, and in the last chapter, a Cause and Effect Diagram (fishbone), and a Pareto Diagram. On the student run chart a red line can be used to show a standard. This gives the

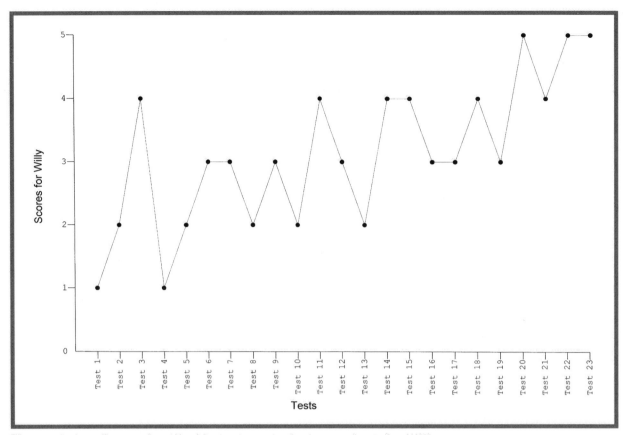

Figure 1.4. Enterprise Weekly tests—student run chart for Willy.

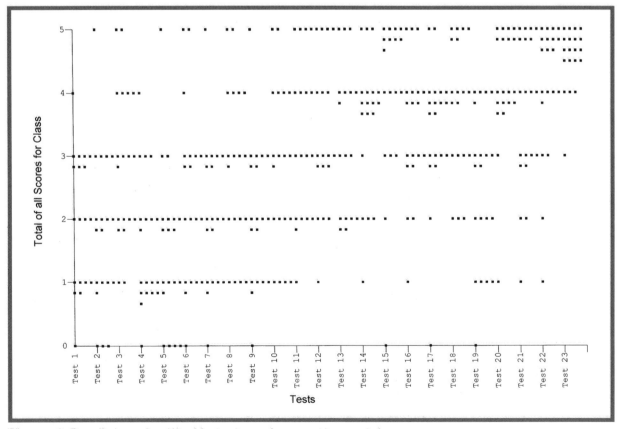

Figure 1.5. Enterprise Weekly tests—class scatter matrix.

teacher, parent, or child a picture of how far or how close the student is from the goal. I think of a line graph like a road map. It tells us where we started, and where we have been; using it, we can predict where we will go. When I look at the Scatter Matrix I am reminded of a Chinese Checkers game. I realize I cannot win the game by moving a few marbles quickly to their places at the other end of the board. I need to use marbles to help other marbles, and I have not won the game until all the marbles have been moved to their positions on the other side.

The Class Action program has a unique feature. An individual student run chart (line graph) can be overlaid onto a class scatter matrix. This gives a picture of how the student is doing in relationship to the class. It is a powerful tool to use with a parent who asks for this information, or who thinks his/her child is doing better or worse than he/she really is (see Figure 1.6).

I am continually learning more ways in which the Class Action program can be used to report information, create student enthusiasm, and monitor growth. One of my recent discoveries was learning how to use the data to indicate student mastery of a subject. In his book, Dr. Jenkins[2] explains how it takes seven weeks to determine if a student knows something, or if his/her

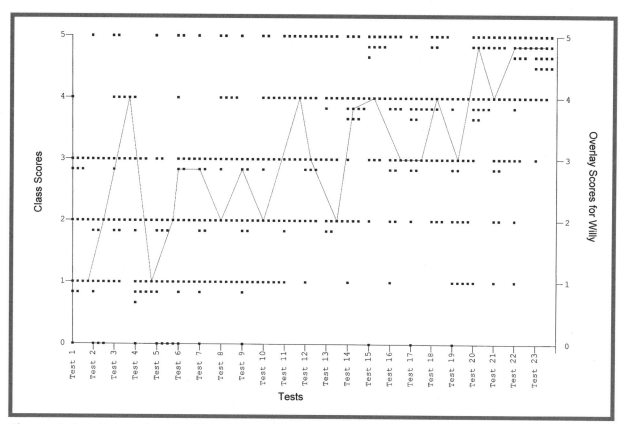

Figure 1.6. Enterprise Weekly tests—class scatter matrix with overlay.

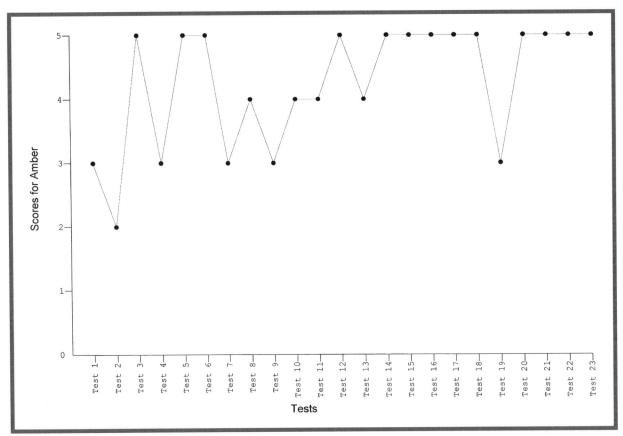

Figure 1.7.   Enterprise Weekly tests—student run chart for Amber Hall. Used with permission.

score is a result of chance. After his mathematical explanation, he says, "Conservative statisticians say that seven weeks of collecting data after a change has been made is necessary to prove whether the growth is due to luck or improvement. Why seven weeks? Because statisticians want to be able to say there is less than a 1 percent chance the growth was caused by good luck. It seems reasonable to accept this advice when students are quizzed once a week in a particular subject. When monthly quizzes are given, however, four months of consistent growth should rule out good luck as the reason for the improvement." With the standard being 4, Amber mastered material from the Enterprise Weekly by week 16 (see Figure 1.7).

Changing grades to standards is just one of the things I have done this year to help students maintain their yearning for learning. Large wall graphs, replicating the Class Action graphs, inform the children about their learning. The use of graphs to track growth is seen as a game, and I know that games are a sure-fire way to guide students toward the learning I want them to be yearning. Besides the game aspect, the children come to realize they are the ones responsible for the tracks the graphs are recording, visible testaments to their learning.

A teacher does not need to have the Class Action computer program to put these ideas to work in the classroom. The wall graphs that guide the learning throughout the year are simple to make using grid paper chosen to the size desired. I used a one-inch grid paper, with each square equaling two points. The maximum points possible on any given assessment was five (explanation of assessments to follow). This, multiplied by the twenty children in my class, gave me the lid—100 class points, making the graph fifty squares tall. Across the bottom of the chart I named each assessment. Sticky dots were used to chart our course. Charts half the size can be made with half-inch grid paper. Each child's own scores can be kept in a regular grade book with the name of the assessment at the top of each column, and the numerical score included in the place where grades are usually kept.

Information and Knowledge

Growth is being tracked in both the areas of information and knowledge. Before continuing, it is important to make a distinction between the two. Dr. Deming divides learning into two categories: information and knowledge. Both are needed to complete the learning circle. As I begin to understand what this division means, I see parallels in many human endeavors. Information is gained as one learns the steps to a dance; knowledge manifests when the steps are used to *become* the dance in all of its fluidity and grace. Information is gained as one learns to shoot a basketball; knowledge manifests when one passes and shoots the ball in clockwork timing during a basketball game. Information is gained as one learns to follow a recipe; knowledge manifests when one creates a culinary masterpiece. In mathematics, information is gained when one learns to compute, measure, identify shapes, and multiply fractions; knowledge manifests when one uses this information to problem solve.

Dr. Deming made another distinction between information and knowledge. He said information is about the past and knowledge is about the future. Dr. Jenkins[2] explains it for the educator. "Dr. Deming defined information as facts about the past. A dictionary is full of information, yet it has no knowledge. Spelling is a subject studied in schools that relates the past to today's youth. It communicates how people in the past agreed to spell particular words. Knowledge, by contrast, is about the future. Writing is a subject studied in schools that is about the future: How can something be written so readers in the future will better understand? Students who become proficient in writing can help create a better future for themselves and others."

I began to think about this. The tools of mathematics are numerals. As with letters, numerals themselves are useless. But it is difficult, if not impossible, to think mathematically without them. Using numerals to combine and take apart numbers in ever endless ways becomes the work of mathematicians. A mathematical language created with numerals is maintained in the library of the past. The language is continually being updated, added to the library. This is the library mathematicians use to create the future. The past is known. Izzy Information—being methodical, predictable—has few surprises. The future is uncertain. King Knowledge holds a surprise at every turn. Molding Izzy Information to its use, King Knowledge creates an atom bomb, or puts a man on the moon.

Committing math facts to memory is a way of acquiring information. Math facts come from the library of the past. The future that math facts were useful for in 1920 is a much different future than the one they are useful for today. As time marches on, information tools adapt to the changes the future beckons.

Even though it is useful to talk about them separately, information and knowledge work in tandem. It doesn't make sense to teach either in isolation. In the rest of this chapter and the next, the reader will be taken through my school year, as I use mathematical information and knowledge to teach, assess, plot scores on charts, and reflect upon what the results are telling me.

The two large line graphs on my classroom wall are enlarged pictures of what shows up on the computer screen when I use the Class Action program. One measures information; the other measures knowledge. I will talk first about the tool I use for measuring information. Following will be an explanation of the wall graph that measures knowledge.

A Tool for Measuring the Acquisition of Information

The Enterprise School District has provided me with a binder that contains 36 weekly quizzes for measuring math information. I use this Enterprise Weekly to plot students' growth from week to week. The problems students see each week come from a bank of problems that have been created to cover many of the math skills students should master before the end of the year at that grade level. Such banks have been created for each grade level in the Enterprise School District. From year to year, skills build upon previous skills, spiraling through the grades. From these banks, questions and problems are randomly placed on 36 weekly tests. In our district, teachers are encouraged to administer one five-question test each week. The questions cover information students are responsible for learning from the beginning to the end of the year. Granted, some of the questions are a review of what has been taught the previous year, but when school begins at a new grade level, many of the questions are strange and unfamiliar to the children. It is important to understand the reasoning behind such an unusual approach. "In current educational practice, teachers seldom clearly state what information is to be known at the end of a course. Most often the course begins with students having only a foggy notion of what they are to learn. Sometimes the assignments are clearly stated, but what is to be learned is not settled in the students' minds. . . . In a 1992 seminar sponsored by the American Association of School Administrators, Dr. Deming suggested a radically different way to manage learning. The steps he suggested follow. Second grade spelling, with 200 words for the year, is used as an example to elaborate the steps.

> "Explain to the students the aim of the course. This advice matches the first of his 14 points for management, which is to maintain constancy of purpose.
>
> "Quiz the students on the square root of the total number of words each week (14 is the approximate square root of 200).
>
> "Each week, randomly select items for the quiz. The students are not told the items prior to the quiz because they will cram, and the teacher will not really know if the students are learning."[2]

What a strange idea! Through the first part of the school year, there is a high probability students will be presented with questions about information the teacher has not yet introduced. I have wondered about this. The beauty of mathematics is in its wonderfully sequential patterns. Random selection seems to make number soup. Isn't this confusing for the students? I have always thought children should learn, and then be tested upon that learning, in a logical, sequential step-by-step progression. However, it is a theory that has some nagging inconsistencies. If this is the way children learn, what about all of the incidental learning that takes place when I teach? Students often pick up information I thought they weren't ready for, information plucked from outside the orderly progression. Dr. Deming said, "A single unexplained failure of a theory requires modification or even abandonment of the theory."[2]

Mathematics can, indeed, be seen as a sequence of pattern upon pattern upon pattern. Although I tend to think in a linear fashion, the brain itself does not function that way. We know that children climb a developmental ladder. Through maturation the petals of the mathematical flowers open for them. Psychologists give general age groups for the stages in which children are able to learn new concepts. The *range* in age is given because children mature at such different rates. Any second grade teacher knows that in a classroom, there are children who understand place value, and children who are not ready to understand it, no matter how many ways it is presented.

My solution has been to try to figure out what each child's readiness is, make groupings of students based upon this determination, and then present what I think they can learn. Can I stop being the God and controller of sequence and trust students' own expanding, flowering, exploding intelligence to take in what they are ready for? Can I trust their innate desire to know, while I help them grope their way to the target? I began finding that I could. I did not throw out teaching the sequential beauty of mathematics, but became open to expanding my idea of what that meant.

"It is theory, not experience, that enables one to learn. The purpose of experience is to validate or challenge a theory. . . . Dr. Deming would have teachers use all their knowledge and the wisdom of master teachers to develop a theory about how to improve student learning. The results shown in the graphs either validate that the theory worked and improved learning occurred, or they challenge the theory."[2]

My new theory came out as a question: Is it useful to randomly test students on materials that many, even with practice, will not master until the end of the year? Even though it is difficult for me to let go of doling out chunks of knowledge in a linear fashion—trying to control what the children's brains can and cannot learn at any given time, I have decided to trust, while I observe. It makes me a little crazy to give up the control, but as I have watched the Enterprise Weekly line go steadily up the graph on the wall, I realize the power in this method of assessment. It is a way for students to see what they didn't know turn into what they do know. I have observed a side bonus. Children stop being so hard on themselves. They realize they don't need to know the answers to everything. What a burden a person is to him/herself and others when he/she thinks he/she must know the answers to everything.

How do I present to the students the idea that they will be seeing new information on the weekly quizzes? They are told right from the beginning that problems will arise they have not learned to do. I say, "You are not expected to know the answers to some of these questions. By the end of the year you will have had experience with them all. Isn't it fun to get a sneak preview? Give each unfamiliar question a try. All of us learn by exploring the new and the unknown. Be glad for any mistakes you make. You are learning!" Through the teaching day, I drive my point home by asking, "What is the best way to learn?" The children chorus, "By making mistakes." They are often thanked for their mistakes. I let them know they have helped me learn what to teach, and they have also helped others who are struggling with the same thing. A surprise bonus comes when the students take the standardized test in the spring. They are more likely to give tough problems a try without feeling discouraged and inept.

The Enterprise Weekly quizzes guide me as to what tools the students need to be able to use: numberlines, hundred charts, base ten blocks, pattern blocks, or rulers. I may not be ready to teach the algorithm to a regrouping problem, but I can encourage students to use a hundred chart to count forward or backward toward an answer. We may break two digit numbers into expanded notation and then add or subtract. I've noticed that as we go over their papers, students have a heightened interest in learning ways to solve those problems with which they are unfamiliar.

The Enterprise Weekly happens to be an effective assessment tool that is available to me. However, these ideas can be used with any assessments that are created. The next chapter covers how to decide what to teach. When the essential concepts have been decided upon, a finite list can be created to include the ideas to be taught. Each idea can be written in a list form, and assessments can be borrowed or written to match this body of knowledge. Then, following Dr. Deming's formula, students can be randomly tested weekly on the number of problems that is the square root of the total number of concepts to be learned. For example, let's say the group of teachers I am working with decides to use the state standards as our body of knowledge. From this we list 100 pieces of information that together create a picture of what we think students should know by the end of the year. We formulate questions that present these concepts in many different ways. The square root of 100 is 10, so from this bank, we randomly choose 10 concepts for our weekly assessment.

A Tool for Measuring the Acquisition of Knowledge

Measuring mathematical information has been done in many ways in many places for many years. It is easy to plot on a graph. On a math fact speed test, Billy got 29 out of 55 correct. The next week, he got 35 correct. His line graph shows improvement. Hooray! However, plotting student growth regarding mathematical knowledge has not been a widespread practice. Under the guidance of Judy Flores, our district mathematics consultant, I chose problem solving questions to use in graphing the children's ability to use their math skills as tools for thinking. Three sources gave me plenty of material for stimulating thought: *Research Ideas for the Classroom: Early Childhood Mathematics*, editor Robert J. Jensen. In this book are research articles written by various authors; *Daily Tune-Ups II*, by Brodie, Irvine, Reak, Roper, Stewart, and Walker; and *Problem-Solving Lessons: Grades 1–6*, by Marilyn Burns. Besides using ideas from these sources, I created similar questions of my own. The Enterprise School District provided me with a rubric for assessing student mastery for being able to think mathematically: 2 points for planning a solution; 2 points for getting an answer with an equation; 1 point for explaining the solution process in writing and/or recording additional equations. Five points are possible. Let's score The Raccoon Problem[4] together. Uzi used a good plan to get a solution; he gets 2 points (see Figure 1.8). He got the correct answer, but there is no equation; he gets 1 point. He has not explained his solution process in writing so his total is 3 points. Can a child

The Raccoon Problem

Four raccoons went down to the lake for a drink. Two got their front feet wet. One got its back feet wet. How many dry feet were there?

Be sure to show . . .
1. How you solved the problem.
2. What the answer is including a number sentence/equation.

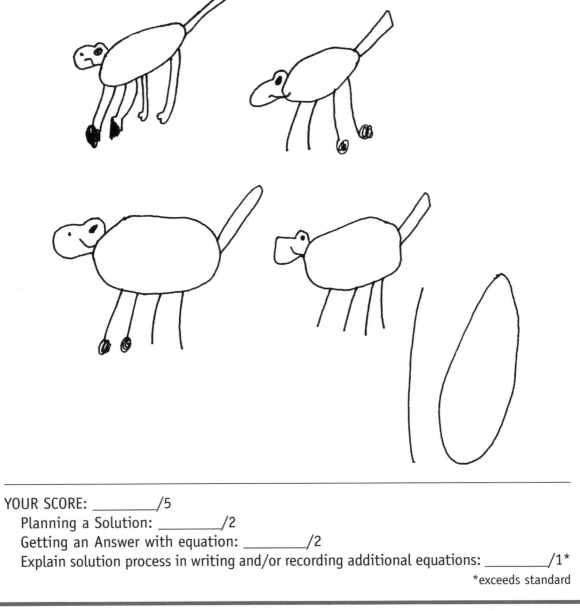

YOUR SCORE: _____/5

   Planning a Solution: _____/2
   Getting an Answer with equation: _____/2
   Explain solution process in writing and/or recording additional equations: _____/1*

*exceeds standard

Figure 1.8.   The raccoon problem. Artwork by Uzi Topete. Used with permission.

Bus Seats

A bus has seats for 50 people. If 23 boys and 19 girls are seated on the bus, how many seats are empty?

Be sure to show . . .
1. How you solved the problem.
2. What the answer is including a number sentence/equation.

$$50 - 23 = 27$$

YOUR SCORE: _____/5
    Planning a Solution: _____/2
    Getting an Answer with equation: _____/2
    Explain solution process in writing and/or recording equations: _____/1*
                                   *exceeds standard

Figure 1.9. Bus seats—Susan's solution.

with the wrong answer get points? Look at Susan's solution for Bus Seats.[4] (see Figure 1.9). She planned a partial solution; she gets 1 point. She has an answer with an equation. Although it is not the equation that gives her the final answer, it is a *correct equation* that leads her in the right direction; she gets another point. Susan has a total score of 2 points. With this same problem, Amber gets 4 points, even though she has an incorrect answer: 2 points for planning a solution. On Getting an Answer with equation, Amber has set up two very good equations. She subtracted incorrectly on the second one; she gets 1 point. She also gets a point for explaining the solution process in writing; the total is 4 points (see Figure 1.10).

By using this rubric, I am able to emphasize the whole thinking process rather than just a correct or incorrect answer. Children can feel good about how well they thought through the problem even when the final answer is wrong. In essence, the rubric says, "I am rewarding you for every bit of thinking you do." My desire is to have students see an error in thinking, and excitedly begin searching for a correct solution, rather than feeling put down and discouraged. The rubric encourages this.

The total of all of the individual problem solving scores becomes a dot on our classroom line graph. As I have pointed out, instead of plotting just answers, I am valuing the children's ability to think. Our line graph is about

---

Bus Seats

A bus has seats for 50 people. If 23 boys and 19 girls are seated on the bus, how many seats are empty?

Be sure to show . . .
1. How you solved the problem.
2. What the answer is including a number sentence/equation.

$$50 - 23 = 27$$
$$27 - 19 = 13$$

I frist toockaway 23
from 50 and then I endid with 27
and then I toockaway 19 from
27 and I endid with 13.
The Eed

YOUR SCORE: _____/5
   Planning a Solution: _____/2
   Getting an Answer with equation: _____/2
   Explain solution process in writing and/or recording equations: _____/1*

*exceeds standard

Figure 1.10.   Bus seats—Amber's solution. Artwork by Amber Hall. Used with permission.

answers *and* about thinking. I wish I had a few simple line graphs like this for plotting my own growth in solving real-life problems. What if when I flubbed up, instead of being mad at myself, I looked excitedly to pinpoint my error in thinking? What if I acknowledged myself for my thought processes? What if I saw that my error actually showed me how to get more quickly to my goal?

Measuring Enthusiasm

I actually have more than two graphs on the wall. There is another I am watching closely. Its title is Math Enthusiasm. If children are hating math, I need to rethink what I am doing. But as I look, I see the enthusiasm graph looks good. It does some up and down stuff, but it is creeping toward the top. To create this graph, the children are given a page in which they fill in the bubble of their choice: I love math; I like math; math is okay; I don't like math; I hate math. Each choice is assigned a number, with I love math getting four points, and I hate math getting zero. In checking with the students once a month, I find different things affecting their attitude toward mathematics. I overheard a child tell a friend he had marked I hate math on his question sheet because he was mad at me for making him sit down during free time. I am reminded that a child's attitude toward learning is coupled with his feelings. Think about it. Isn't it true for any of us?

In another case, Nick, who is a capable math student, began to compare himself to Chris, who catches onto things and works more quickly than he does causing Nick to think he was dumb. His rating of enthusiasm for himself went from I love math to I hate math. Seeing this, I was able to talk with Nick about his feelings. We talked about what an intelligent boy he is, about how all children learn and work in different ways. I was able to point out to Nick some of his unique ways of thinking that are like no other. Clearly, how children feel about themselves affects how they feel about the subjects they are learning.

A few weeks after I had my visit with Nick, he shyly handed me a folded piece of paper with my name on it. I opened it up and read: "I am starting to LIKE—love math agen." I burst out laughing in the pleasure of the moment. Nick laughed, too. Because he had been honest enough to mark I hate math, I knew this note was also an expression of an honest feeling. The graph is a map showing Nick's feelings about math (see Figure 11.1).

Now that the reader has had a cursory view of how I use data collection to guide my teaching, a more detailed look is needed. The meat of my math program is all of the things I do that improve learning so that when the weekly assessments are given, we can see growth. My linear view would have me talk about the math skills first, because it is the order in which I have always thought mathematics should be taught: math skills first, then problem solving.

I am realizing now that this is one of those unquestioned axioms of truth, sort of like the Earth is flat, that limits the opportunities I give children. The line of thinking is, "I can teach the children to problem solve as soon as I give them the skills to do it." What if I were to teach them both at the same time right from day one? Students would have to come up with their own ways of thinking in areas where they did not yet have strategies. I decided to test this idea as a theory. Here we are with one of Deming's axioms of truth: Improvement comes from testing theories. The plan would be to see what happened if learning math skills and learning to problem solve could be successfully taught concurrently. I would collect data, talk with the children, think about

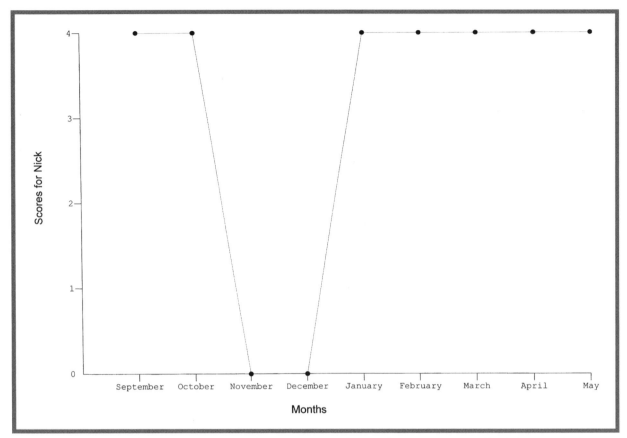

Figure 1.11.   Math enthusiasm student run chart for Nick Carreiro. Used with permission.

what I saw, change, and try again. In this way, if something was not working, I would have the feedback I needed to govern my change. If it worked well, I could do more of the same for further verification and improvement.

Problem solving is multidimensional mathematics. As children plan their strategies and make their computations, they practice math skills, while at the same time discover what they are for. Having a purpose increases the desire for mastery. Working with both math computation and problem solving, right from the beginning of the year, seemed to make sense.

Here was the question for which I set out in search of an answer: Without being given special preparation, can children construct their own ways to solve problems? I knew that in September the children would not have had much experience in using problem solving strategies, but they would learn as they went, and always, mathematical thinking aids and tools would be made available: play money, rulers, numberlines, hundred charts, pattern blocks, rainbow tiles, base ten blocks, and calculators. ". . . Abstract thought is not ordinarily characteristic of the students with whom we deal in grades K–4. Obviously then, our students—even those highly motivated—will encounter developmental obstacles in attempting to reason logically, deductively, and symbolically. To overcome these limitations, we simply must rely on concrete representations of problems. The pupils may then manipulate the physical model to conceptualize and resolve the problem. As the pupil's mental development progresses, a semi-concrete approach with pictures and diagrams can complement and supplement the use of purely concrete representations."[4]

Jumping into Problem Solving with Both Feet

This was my first year to use problem solving as a cornerstone of my math program. When I looked at the problems I had chosen, they looked too difficult for most of my students to solve. I wondered if I should go ahead and use them. Would the children get discouraged? Had I made the right decision in beginning so soon with it? Here was my own problem to solve, with no rubric to guide me. To figure out where to begin, one must begin, so I began.

It was October. The children sat silently at their desks and worked their first problem. Rides at the Park: 12 rides at the park were scary. 13 rides were funny. 15 were silly. 7 were for little children. How many rides were not scary?[5] It was a disaster! They whined that it was too hard; they didn't understand what to do. Their body language showed me they were intimidated and discouraged. Math enthusiasm was way down; mine, too. With this problem I now had a score for the first point on the problem solving graph, but the price for that point was negative feelings and discouragement. Students felt I had given them a task that was impossible for them to solve. Why was I doing this to them? (See Figure 1.12 for a paper that received 0 points, but as one can see, Jared gave it his best.)

I decided if the children could work together they would have more fun, and possibly, they would find it easier to come up with correct solutions. For the next two problems they worked together. I could not check the papers for a personal student score, which would then give me a class total for the graph, but everyone seemed happier. It made me feel good to see everyone happier, but there was no accountability, no system for checking progress; how could I adjust the rudder, if I couldn't see our present position? I needed more specific information than just observing the children, and then looking at their collaborated papers when they were finished.

For the next two problems, I went back to having the students work in isolation. I scored the papers and put the results on the graph. We were all discouraged again. What could we do? How could I get an individual score and still allow the children to work together to solve their problems? What I did next, is integral to a Deming system; I talked with others. Talking with others is such a simple act, but I must remind myself to do it. I spend hours alone in my classroom with the children, and I start thinking I must figure everything out myself. Talking with others gives me new ideas to try; or the talking

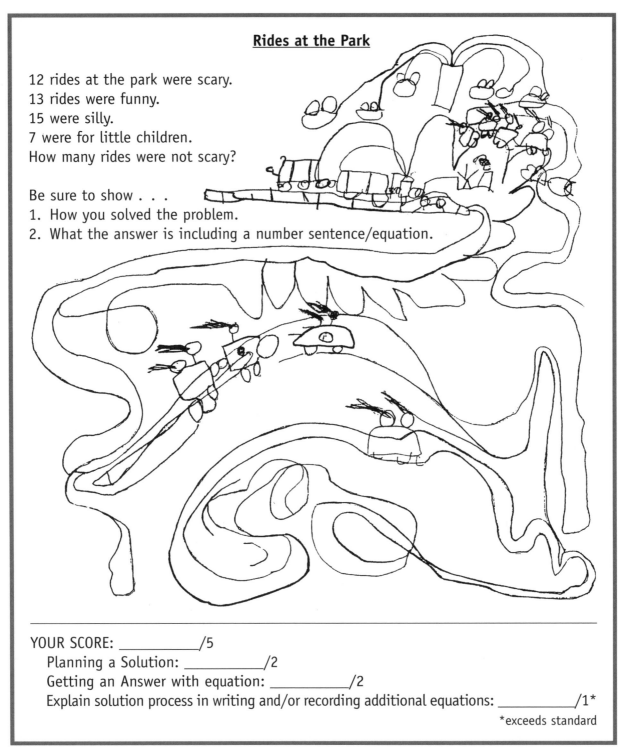

Rides at the Park

12 rides at the park were scary.
13 rides were funny.
15 were silly.
7 were for little children.
How many rides were not scary?

Be sure to show . . .
1. How you solved the problem.
2. What the answer is including a number sentence/equation.

YOUR SCORE: _____/5
   Planning a Solution: _____/2
   Getting an Answer with equation: _____/2
   Explain solution process in writing and/or recording additional equations: _____/1*
                                                                    *exceeds standard

Figure 1.12.   Rides at the park. Artwork by Jared Lofton. Used with permission.

becomes a catalyst that reconnects my brain paths to information I already have. It has been my observation with myself and with the children I teach that if connecting of information from one part of the brain to another does not take place, the information might as well not be there. It sits, making no contribution to the solution of a problem. All the information I needed to solve this problem was in my own brain, sitting. A bridge was made when I talked with others.

One of my colleagues shared that the Japanese, when doing problem solving, work on the problem first by themselves. When each person's thoughts have had a chance to form, *then* they work in groups to come up with a solution. This information reminded me of the problem solving I did one summer as a part of the California Mathematics Project, and from this, a solution emerged.

It is an interesting process, this business of having solutions emerge. The key is commitment to the ideal that is established. Without the commitment, I am too likely to turn around and go down another road, one with fewer roadblocks; or I may just give up. How connected am I to my aim? How strong is my desire? I have seen how tenaciously I can hold onto something that has been born out of a strong desire. In this case, the desire was mixed with curiosity: Could I use data to show how we were doing, and at the same time maintain student enthusiasm?

On a five-point scale, I give the following collaborative solution a five. It does not matter that it is was not my idea. I do not need to create first-ever solutions to my problems; I just need to find what works and do it.

When I problem solved as part of the California Math Project, my fellow students and I worked together, usually in groups of four, to solve The Lunch Problem, named because each day a different group ate its lunch while it first worked the problem for that day, and then planned how to present it to the rest of us after we returned from lunch. To begin the afternoon problem, we sat silently in our groups of four and each took two minutes, to read and think through how to solve it. Only after this quiet time did we begin working together. The thinking time was key to the process. At any time, we could use the manipulative tools available in the room. The use of unique and varied strategies were encouraged and all points of view were respected.

At first I felt intimidated as I sat in groups of high school teachers with their fancy algebra equations. But I had no choice. I was told to think. I worked the problems in the only ways I knew, and in some ways I'd never thought of before. Because all thought processes were honored, I came to gain a deeper confidence in my own ability to think. I began to look forward to each new problem.

As the days went along, I worked with different people. Our groups were formed as we entered the room. The presenting group was encouraged to come up with clever ways to group us. Here are a few of the ways we were grouped that I remember. (1) We were each handed a piece of junk as we entered the room—a key, a button, a pencil. We had to find three others with junk that had a similar attribute, a decision to be made among ourselves. (2) We matched our candy pieces, geometric shapes, playing cards, or linkercube color. (3) We were given pictures of things that go on a pizza—mushrooms or tomatoes and we found those like our own. (4) We took off our right shoe. The shoes were put into a pile. We took someone else's shoe, found out who it belonged to, then found another pair of people to link up with. (5) Four ways to tell the same time were passed out—five fifteen, 5:15, a clock face with the time, a Roman numeral face with the time. We were each to find others with the same time. (6) We matched a part of a comic strip to the other three parts. (7) We each drew out an addition combination and sat where the answer was.

Before we began to work, we were assigned jobs. One person in the group was the recorder and one the reporter. These jobs were assigned randomly so that all people eventually ended up getting to do both jobs. When it was time to share, the reporter brought the drawings, equations, and written explanations of his/her group to the overhead projector where those seated could see how another group had solved the problem. During the explanation, if someone in the audience saw something that was incorrect or that bothered them, they could ask questions, or challenge the thinking. Listeners had as much responsibility as the presenters.

Were my second-grade children too young to follow a similar process? There was only one way to find out. Try it. The next problem read: How Many Legs? There are 3 cows. Each cow has 4 legs. How many legs are there in all?[6] They had worked similar problems in first grade. It would be a good one to begin with. Standing file folders were used to make little study carrels where each child could work by him/herself. The room was silent as the children intently drew cows and scratched out equations. Everyone was done in 15–20 minutes and I collected the papers. When I scored them, the only marks I made were in my own roll book. The next day the unmarked papers were returned and the new process was explained. Children already sat in groups of four in the room. We used these groupings and the process began. For a few mini-moments, as the children conversed, I thought I had died and gone to teacher heaven. This looked to me like a storybook teacher's dream come true: Children excitedly talking to each other with their opinions about how to solve a problem. However, when it came time for presentations at the overhead projector, I realized I had some adjusting to do. It was obvious I had not made it clear that all members of the group needed to agree to the solution before it was recorded. Cathy's own paper had a correct solution, but the recorder, Mack, had transferred his own solution to the overhead transparency. As the reporter, Cathy went into a state of confusion trying to explain Mack's solution: "We added $3 + 4 = 7$ to get seven legs." I asked which cows would have to go without legs. I saw Cathy's confusion mounting, so I asked this group, Cathy, Mack, and Cody, to go down the hallway to work on it some more while others were making their presentations. As they walked away, I heard Cathy yelling, "It's twelve. It's twelve. I

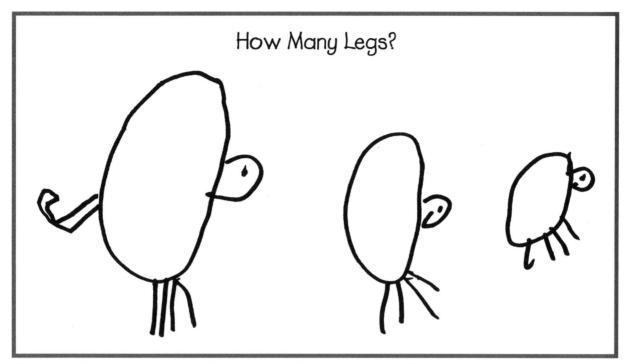

How Many Legs?

Figure 1.13.   How many legs? Artwork by Cody Gustafson. Used with permission.

know it's twelve." But no one listens to Cathy. She yells a lot. Mack grabbed a calculator on his way out. Later when I checked on them, they still had no solution, and Cathy was continuing to rant, "It's twelve. It's twelve!" I asked, "Would you like to hear how some others solved the problem?" Yes, they would. "Some groups just drew the cows and counted the legs." With Cathy looking the epitome of smug, they soon emerged with their drawing (see Figure 1.13).

Was this first group process successful? I looked at how the students felt about it. Even though I was pushing some children right up to the edge of their patience and ability to focus, most of them were fully engaged. The room was alive with anticipation and excitement. I had some fine tuning to do, but I had begun. Beginning something is always the most difficult, especially if I want everything to run smoothly, no hitches, all the kinks worked out the first time—and I always do want that. But like with any learning, I can't adjust what I'm doing to make it better until I do it.

The next time we worked in groups to create presentations, I experienced the real power of group problem solving. Individually, only one child in the class had a correct solution to the following problem: A bus has seats for 50 people. If 23 boys and 19 girls are seated on the bus, how many seats are empty?[5] Besides the child that had the correct solution, there were three others who had equations that would have led to a correct solution, if they had added or subtracted correctly. When the five groups of four each came to the overhead projector to present their solutions, all but one of the groups had come up with a correct solution, each finding different ways to do it (see Figures 1.14–1.17). I

23 + 19 = 42 + 8 = 50

23 + 19 = 42 and
We know 8 more is 50
so the anser is 8

By   Nick
Ray   Cody   Lai

Figure 1.14.   Group solution #1.

50 − 23 = 27
27 − 19 = 8   by Amber Nowell Hall   Amber G.   Zach

We uoosed a cakylater.

Figure 1.15.   Group solution #2. Artwork by Amber Hall, Zach Powers, and Amber Guillaume. Used with permission.

Figure 1.16. Group solution #3. Artwork by Terica Carpenter, Kelly Hill, and Katie Minks. Used with permission.

Figure 1.17. Group solution #4. Artwork by Jared Lofton, Ray Salazar, and Markyy Holmes. Used with permission.

was amazed! They had learned from each other. I knew this because, although I encouraged them to talk to each other, to stay with their problem until everyone agreed, and to use whatever tools they needed, I gave no direct help. I felt the children were learning more from interaction with each other than they were from watching me as I showed them how to do it on the overhead projector, a method I had used on the earlier problems that had no group participation.

Studying the Data

As we placed our dots on the big problem solving graph, the lines drawn between them began to create what looked like tall volcanoes and deep chasms. Were the children building upon what they had learned from the last problem so as to do better on the next? In the process of creating continual improvement toward a goal, Dr. Deming states that variation is the enemy. Looking at the graph, it seemed our enemy was rather large. Using Dr. Jenkins'[2] book, I reviewed Dr. Deming's philosophy on variation.

He says variation cannot be eliminated. However, understanding its source can be helpful. There are two ways variation manifests itself. One kind of variation is caused by probability that exists as a universal reality; the other is caused by unusual events. If half of my class gets the chicken pox and stays home for two weeks, a huge variation appears on my problem solving graph, since the graph is being created with class totals, and half of my class isn't there. This is a special-cause variation. The other kind, common cause variation, is built into the system, or just luck. My goal is to tighten the distance between the tops of the mountains and the bottoms of the valleys, while at the same time having the line crawl upward. When I see a big jump in the line either up or down, I must decide if there is a special reason for the jump, or if it is a part of the expected variation—expected, because variation *is*.

Children are directly affected by their immediate world. It doesn't hurt to think outside the box when brainstorming special reasons for variation. The week after Halloween when everyone is on a sugar high might easily affect the chart scores. The children saw a noticeable change after one of the class' best math students moved. Derek had been adding five points into the total score on almost every assessment. The class was particularly smug the day their score on the Enterprise Weekly beat the previous high score, this time without Derek's five added points. Also, much can be learned from watching the student run charts of individual children. Sarah lives with her grandmother. When her mother came to visit for a few days, Sarah was walking on clouds. Her student run chart gives us a record of the time right after her mother left again (see Figure 1.18).

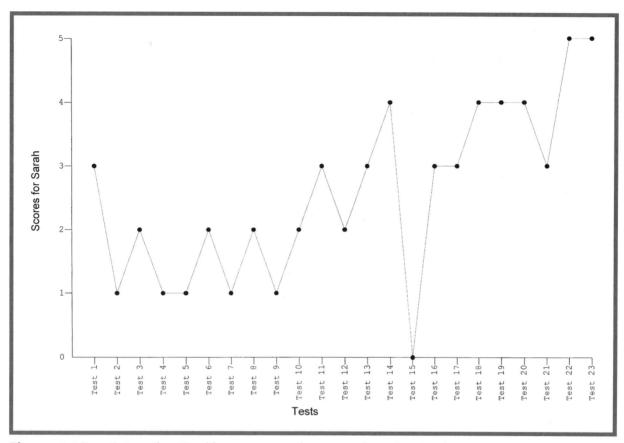

Figure 1.18.   Enterprise Weekly tests—student run chart for Sarah.

An obvious added benefit found in this careful looking, is that besides getting to know about their own progress, students are having the experience of analyzing data—looking at the dots and lines, knowing meaning can be extracted from them.

After looking at the wide variety of problem solving questions the children were being asked to do, I decided a wide variation on the graph could be expected. The class as a whole scored well on Ducks in a Puddle: There are 5 ducks playing in a puddle. Soon 9 more ducks come to play. Then 5 more ducks come. How many ducks are there in all?[6] The children were familiar with this kind of problem. On the other hand, the score was way down on At the Beach: Sai picked up 29 shells on the beach. He put back 16 of them. Then he picked up 14 bright pebbles. How many shells and pebbles did he have?[6]

With the differences in the problems, how could the chart show anything that would help us adjust our rudder to get going in a straighter line toward the top? I had a feeling we were learning, but how could I know for sure? A few weeks after our scores bombed on At the Beach, the children were presented a similar problem—Balloons for Sale: The balloon man had 65 balloons.

He sold 19 of them. He pumped up 15 more. How many balloons did he have then?[6] This was a chance to compare apples with apples.

The children set up their study carrels and went about quietly working the problem, each in his/her own way. My heart soared as I saw all the different ways of thinking. A few were counting on their fingers. A couple of children were using a hundred chart, another a calculator. One child was drawing balloons—lots and lots of balloons. You could almost *hear* the intense thinking going on.

My heart soared again when I checked the papers and saw that we had improved. The next day the children watched as I placed the sticky dot on the chart 16 points above At the Beach. The smiles on their faces showed they were delighted and impressed with themselves. We needed to gather more evidence, but here was an indication that we *had* learned.

I continued to fine tune. The simplest way I know to do this is by looking at what is not working. Sometimes participants in a group did not interact well. In the classroom, students sit where they want, in groups of four, unless a reason arises for me to change the seating. On the first day of each new month students may voluntarily change their seating. We worked in problem solving groups right after the December desk scramble. A couple of groups spent most of their time arguing about who would be reporter and who would be recorder. Was it luck that the groups before had worked so well together?

I talked with Judy Flores, a math consultant in the district, who shed some light on my befuddlement. She said it is better for students to stay with a group for a good while, because as students inside the group get used to working with each other, all the preliminary shuffling and jockeying for position disappears. Students get comfortable talking through their problems and they begin a history of working together.

I thought about this for myself. I work with the same group of teachers throughout the year. When we meet to plan curriculum, or to look at assessment, we get right down to business. With a limited amount of time to meet, we do not spend our time jockeying for position. Instead, our energy is focused upon what must be accomplished before our time is up. Jamie usually takes the lead, and moves things along. Mimi asks many clarifying questions. I help us stay focused on where we're going. An important contribution comes from Patty, who has a wonderful sense of humor. She keeps us lighthearted and laughing. We're used to the way our working puzzle fits together. Thinking about this, I realized that stirring the pot in the classroom gave students a chance to deal with change, which is a valuable skill, but it did not immediately enhance cooperation of groups for problem solving.

Judy made another suggestion: Older students are more able to work in groups of four. Second grade students might be more successful with groups of two. When I looked to see if the smaller groups was something I wanted to do,

Plus-Delta Chart Question: How do you feel about your partner problem solving task?	
+	Δ
We didn't fight. We both did the picture. I got to teach. I taught Warren. He taught me.	I didn't get to do anything. I didn't get to draw. Jerry wouldn't listen to me.

Figure 1.19. Plus-Delta chart.

I realized putting children in pairs would double the number of presentations. I could not require rapt attention to an hour of presentations. Thinking back to the Math Project, I remembered that, usually, all groups did not do a problem solving presentation. After each presentation, the instructor would ask, "Is there someone who has a different way of solving the problem?" This encouraged us to value unique thinking, kept us from tiring, and increased our repertoire of strategies. In doing this with my own students, I would need to be sure to acknowledge groups who did not get to present, so they would feel validated for their hard work. Another surefire way to cut down on disagreement in the groups is a strategy any teacher knows: assign the jobs. Four different colors of M&Ms in a cup at each table. One of the group members holds the cup high enough the candy cannot be seen. Each child draws an M&M and each color represents a job: reporter, recorder, reporter helper, recorder helper.

Periodically, and maybe it would be best to do this each time, I do a Plus-Delta Chart[7] at the end of the cooperative group time, asking the children how well their group worked together. A Plus-Delta Chart is a way to get feedback from students. A simple piece of paper with two columns shows a Plus side where students list what they like, and a Delta side where they contribute the things they don't like, or that need to change. I used the Plus-Delta Chart one day after I had put students in groups of two. This day I paired students who understood how to solve the problem with partners who did not. The students were not aware of my criteria for pairing, so when I asked them to be willing to teach and learn from each other, all partners were eager. When we filled out the Plus-Delta Chart, these were some of the comments contributed for the Plus side: "We didn't fight." "Meghan helped me with directions." "We both did the picture." "I got to teach." "I taught Warren. He taught me." On the Delta side I got: "I didn't get to do anything." "I didn't get to draw the picture." "Jerry wouldn't listen to me." (see Figure 1.19). We talked about ways we could work through these snags.

The ultimate way to fine tune my teaching is to get direct feedback from individual students. I like sitting one-on-one with the intent of listening as the child tells me what he/she was thinking while solving a problem. Besides giving me valued information, which I can then use to adjust my teaching, a student feels acknowledged. I think about how I feel in a similar circumstance. When someone asks me a question, and then really listens to what I have to say, I feel truly acknowledged. Children feel the same.

Lai Fin

Lai Fin was a child I needed to spend time talking with and listening to. In the area of math computation, he was one of my best students, but each problem solving paper showed inappropriate computations for what was being asked. This was my second year with Lai Fin. At the beginning of first grade, the only thing I could understand him say was, "Can I go to the bathroom?" Lai Fin is Mien, and no English is spoken in his home. What courage it must have taken for him to come to school where everyone spoke a foreign language!

Lai Fin had a strong desire to succeed, and through his first grade year and now into his second, was learning the English language rapidly. Even so, he was still at a great disadvantage when trying to understand the problem solving questions. As we began each new problem I would ask him, "Do you understand?" I realize now that he would always answer yes because he did not want to appear stupid or foolish.

One day I sat with him to go over a problem with which he had done poorly: Snow White and the 7 dwarfs came walking along. Each one had 3 apples. How many apples did they have in all?[6] At the end of our talking about the problem, Lai Fin drew a picture and solved the problem correctly (see Figure 1.20)

I asked if the little four-legged creatures were dwarfs? Lai said, "Yes." I asked if he knew what a dwarf was. "No." Did he know the story of *Snow White and the Seven Dwarfs?* "No." Poor Lai Fin had had to wade through something called dwarfs and White and Snow to get to the essence of the problem to be solved! We talked again about what a good math student he was. Why was he doing poorly on the problem solving? I told him I thought it was because he did not understand what was being asked. We agreed he would do no more problems without being sure he understood what to do.

The next problem was Sharing 50 Cents.[4] Lai Fin was one of the few children in the class to get all 5 points! He made sure he knew what to do. Then he methodically divided up pennies to get his answer (see Figure 1.21).

$$3 + 3 + 3 + 3 + 3 + 3 + 3 + 3 = 24$$

Figure 1.20.   The seven dwarfs. Artwork by Lai Fin Saelee. Used with permission.

From that point on, Lai Fin's problem solving took a dramatic upswing (see Figure 1.22). What if I had not taken the time to talk with him? And a better question is, why didn't I do it sooner?

Uzi

Uzi was another child with a language disadvantage. His father spoke English, but his mother did not. Uzi was bilingual, with Spanish being his dominant language. He came to my class after the school year began. A big, slow-moving boy, he spoke in a slurred English that was hard to understand. From the moment he stepped into my classroom, we were in a constant tug-o-war. Resist, resist, resist. One day I observed him in his problem solving group. Three children were poring over the problem. Uzi was drumming the desk with two rulers. This was not unusual. It was clear that Uzi was not relating to the other children, nor was he engaged in his learning. When I pulled him into the spotlight of my thoughts, the problem began to solve itself. He had shown me he could read and spell quite well. And during a small group math lesson, I saw that he caught on quickly to new concepts. A shy little smile showed he was pleased when I commented on how easily he had learned what I was teaching. I suspected that behind the resistance there was a bright and inquiring mind.

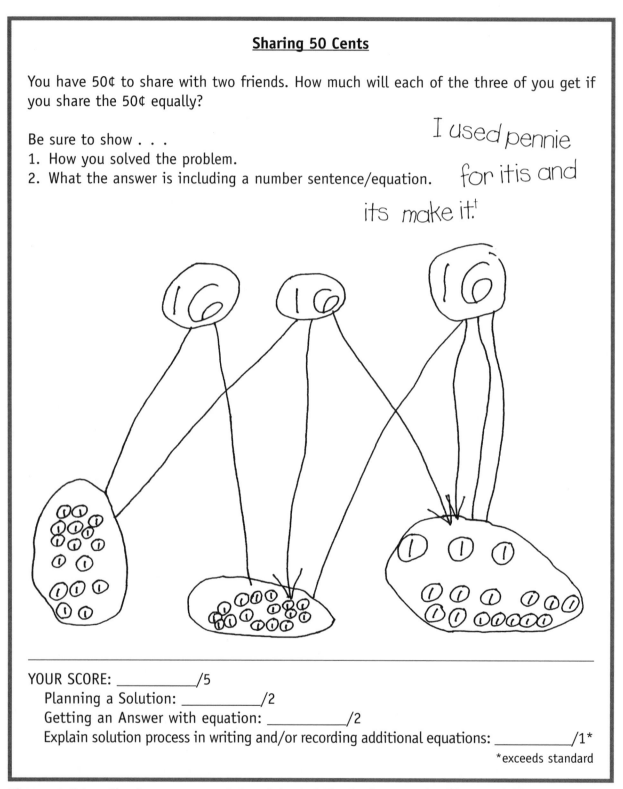

Sharing 50 Cents

You have 50¢ to share with two friends. How much will each of the three of you get if you share the 50¢ equally?

Be sure to show . . .
1. How you solved the problem.
2. What the answer is including a number sentence/equation.

I used pennie for itis and its make it.†

YOUR SCORE: _____/5
   Planning a Solution: _____/2
   Getting an Answer with equation: _____/2
   Explain solution process in writing and/or recording additional equations: _____/1*
                                                                          *exceeds standard

Figure 1.21.   Sharing 50 cents. Artwork by Lai Fin Saelee. Used with permission.
†This wording is Lai Fin's struggle with expressing himself in English. When I talked with him, his meaning was "I used pennies, and by using pictures I figured it out."–Author's note.

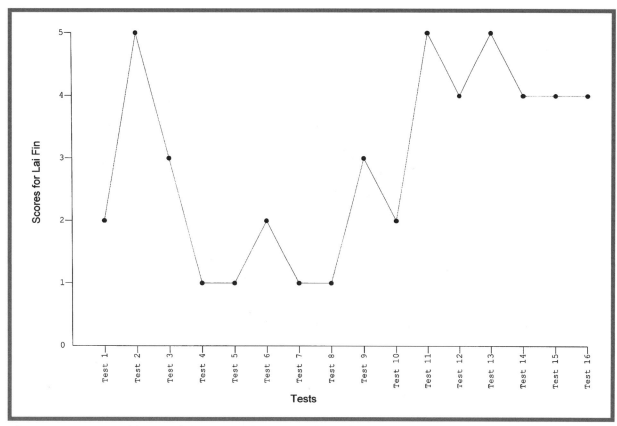

Figure 1.22.   Problem solving—student run chart for Lai Fin Saelee. Used with permission.

The following week the class was presented with the aforementioned problem of Balloons for Sale. After the children had erected their file folder walls and were working, I walked over and whispered in Uzi's ear. "This is a problem I know you can do. Look and think carefully. You can do this one." I wandered around the room and then came by Uzi's desk again. He was busily drawing balloons, stopping to count and recount. As most others were completing their work, I heard him whisper excitedly to himself, "It's sixty one. I *know* it's sixty one." When all the papers were in, I perused them. Uzi was one of the few with a correct answer. He had no equation or explanation for his thinking, but he had a correct solution. The next day when the groups worked on the problem, Uzi was acknowledged by his group and later by me for his good thinking. He was hooked.

A week later, Uzi again worked carefully on the assignment: The Beaver Problem. There were five beavers making a dam. Three beavers got their front paws muddy. One beaver got his back paws muddy. How many dry paws were there? I encouraged him to write an equation to go with the problem. This would give more points, I said. Uzi solved the problem correctly with a drawing , but accompanying it was this equation: $10 + 10 + 10 + 10 + 10 + 10 + 10 + 10 + 5 = 85$ My, my! Where had that come from?

It was time for me to listen. I sat with Uzi as he explained his thinking. He told me how he had figured out the problem, but did not mention the equation, so I asked. "Tell me how this equation ties in with your problem." "It doesn't," he said. "You told me to write an equation. There's an equation." I had to stifle a laugh. What if I hadn't asked? I could now stop worrying about Uzi's reasoning skills. He put the equation there because I asked him to write an equation. He must have been thinking, "I don't know what she wants that equation for, but she's made such a big deal, I'll give it to her to make her happy." Feedback provided me with what I needed to do to guide Uzi to his next understanding, that of being able to write an equation—not just any equation, but one that fit the problem.

Uzi gave me a true indication of his feelings about problem solving one afternoon just before dismissal time. I was writing on the Plus-Delta Chart, getting responses from the children about things they enjoyed during the day, and things that didn't go so well. On the plus side, I recorded such statements as: "I liked P.E." "Brad shared his cookies with me." Uzi contributed: "We got to do problem solving." Music to my ears!

Uzi began greeting me each morning, and telling me good-bye at the end of the day. When we did a creative writing lesson, using similes, he wrote: "As hot as the fire, as cold as the iceberg. As big as a T. rex, as fun as being in Mrs. Ayres' class." I thought I had Uzi hooked? He had me hooked!

If Something Is Not Working, Do Something Else

So with all this listening and all this doing, was I helping the children to become great problem solvers? January was here and even though we had evidence of some learning, the graph as a whole looked pathetic. And there it was, huge—you couldn't miss it—up there for the whole world to see. Why did I have to make it so big? Look, parents. Look, teachers. Look, Mr. Principal. Mrs. Ayres can't help the class find their way out of the mire (see Figure 1.23).

I recently talked with a new teacher at our school. She told me how much she loved the children; how much she loved to teach. But she was experiencing a heaviness from the constant pressure of needing to show results. She wondered if she would be able to prove herself as worthy of her position by the end of the school year. Did other teachers feel this? I assured her we did.

I posed the question to myself: How do I, as a teacher, live with this constant pressure to produce results? In this present situation, do I compromise the integrity of my teaching as a trade-off for being able to show the principal, parents, and other teachers that I am doing better than I really am? With easier problems the line on the graph would go up. But then it would be an empty, pointless line. Behind any piece of paper, one wants substance. This is true of a

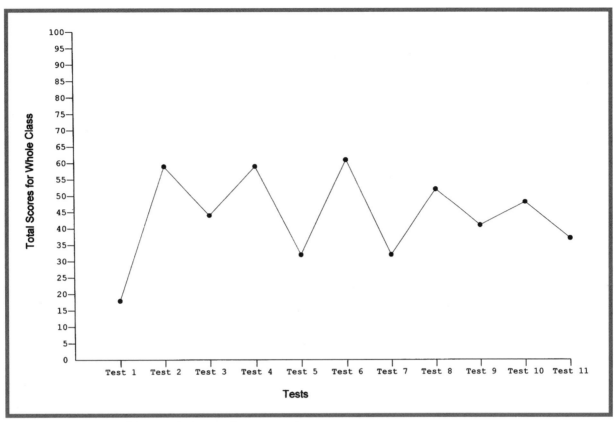

Figure 1.23. Problem solving—class run chart.

dollar bill, a fancy certificate, or a wall graph. The integrity of the graph must be maintained. The joy in teaching lies with achieving the goal of student improvement. I become enlivened by seeing the children begin to believe in themselves, by seeing them grow in their excitement to learn. The pressure of needing to produce results could get me to abandon what I am doing before I have had time to discover how to make it successful. To do this would be what Dr. Deming calls tampering.[2] Patience, trust, and support of long term goals are replaced with the desire for a quick fix. If, in the process of doing my learning experiments, I change plans each time I see the graph go down, I am tampering. If I quit before I have given it enough time, I am tampering.

Had I given my plan enough time? Yes. Seven weeks of data was enough. It was time to rethink. ". . . The results shown in the graphs either validate that the theory worked and improved learning occurred, or they challenge the theory."[2] My theory had been challenged. Although there were individuals in the class who were showing personal growth in problem solving, the class as a whole was not showing they could come up with strategies of their own to be successful with the problems. The students had already done more problem solving than any second-grade class of mine had ever done before. Here would be the time to take the graph down and say, "Well, there's the end of that unit." But I was not willing to abandon the project. I'd seen Uzi get hooked on a problem, and not want to let it go, even to have recess. I was hooked, too. Recess could wait. I was determined to make that line go up.

In my personal life, before I had ever heard of Dr. Deming, I had enjoyed living by the simple axiom: "If what I'm doing is not working, do something else." It kept me from blaming and complaining. The trick was to be aware enough to notice when something was not working. In this case, I knew. How? The erratic, jagged graph line that went across rather than up. The beauty of having a display of my data was that I could easily see whether or not something was working. It wasn't. With the days counting down to the end of the year, it would not be a good strategy to just hope things got better. Something else needed to be done. On a train, this would be called laying new track. In the Plan, Do, Study, Act model, this would be called making a new plan.

The PDSA model was developed by Dr. Walter Shewhart, a statistician, and one of Dr. Deming's working companions. Dr. Deming said Shewart's genius was in recognizing when to act and when to leave a process alone. Shewhart's theories of quality control became the basis of Dr. Deming's own work.[8] As applied to education, the PDSA model can be stated like this: *Plan:* (1) Determine what the current system is producing. (2) Analyze the data for causes. (3) Envision improvement in my classroom system. *Do:* (4) Decide which improvement theory to attempt. (5) Implement the theory. *Study:* (6) Study the results of the experiment. *Act:* (7) Establish the changes which resulted in improvements.[9] (8) Start over.

It was time to create a new improvement theory. What was hidden in those jagged lines that refused to ascend? Peggy McLean, an acclaimed math educator, was scheduled to be at our school as part of an ongoing guidance with our school's Program Quality Review and would be observing in my classroom. The line on the graph had taken a real nose-dive as a result of our last problem. Should I ask this revered person for help? I swallowed my pride and invited Ms. McLean to watch as the students worked in their problem solving groups. I asked her to shed some light on my perplexing dilemma. The nose-dive problem the children were working on read: Sharing 50 Cents. You have 50 cents to share with two friends. How much will each of the three of you get if you share the 50 cents equally?[3] Peggy walked around and talked to students as they worked. Some groups were engaged in meaningful discussion; some didn't have a clue what they were doing. When the students had gone to recess, Peggy said some things that shifted my view and caused me to begin a new plan of action. She said the problem was too hard. The children needed to be given manipulative activities that would lead up to being able to do the kind of problem I had given them to work on. Bingo! My plan had been to have children come up with their own ways to work through the problems. To give students activities that were similar to the problem they would be doing seemed to be taking away their own chance to figure and think. The reality was they were getting plenty such chances, and not being successful. Most were too young—they hadn't been around in this world long enough to have the sorts of information needed to extract solutions to these problems. Peggy gave me a new way to think. I know that ultimately everything I do is a result of what I think. If I want to change what I do, I must change what I think. When Peggy said the problem was too hard, I thought, "Duh! Of course it's too hard. I have done no division activities with the children, and even though we've worked with coin values, many do not have facility putting together and taking apart money amounts."

I created a new improvement theory, again phrased as a question: If students are given practice activities that lead up to solving a problem, will they be able to transfer the knowledge and successfully come up with a solution when given that problem? I decided to write my own problems that were similar to ones we had already done. But before doing the problem, students would get a chance to practice in a manipulative setting.

I talked to the children about the new plan. We looked at the graph. I explained that it was possible I was not giving them enough information to have them succeed with their problem solving. I told them we would be practicing before we worked the next problem.

The new problem the students would be getting ready for was this: Uzi's Marbles. Uzi had 25 marbles. He gave his friend Cathy 9 of his marbles. The next day Nick brought marbles to school. Nick gave Uzi 16 of them. Now how many marbles did Uzi have? It was similar to the Rocks and Shells problem. Rather than just hope we would do better, we would practice first!

To get ready for Uzi's Marbles, I set up math stations using buttons, keys, rocks and shells, and watermelon seeds (navy beans sprayed black on one side—used for what is called a Work job box). The students were to plug in their own numbers, explain their thinking in writing, and give equations.

Buttons

Meghan had _____ buttons on her dress. She lost _____ buttons. Her mother gave her _____ new buttons. How many buttons does she have now?

Shells and Rocks

Jared had _____ rocks. He gave _____ rocks to Mike. Then Amber gave Jared _____ shells. How many rocks and shells does Jared have?

Keys

The custodian had _____ keys. He lost _____ keys. The principal gave him _____ keys. Then how many keys did he have?

Watermelon Seeds

Cathy had _____ watermelon seeds. She spit out _____ in a spitting contest. Cody gave her _____ more seeds. How many did she have?

The students showed their usual enthusiasm when given meaningful activities to complete. We all had fun. A few days later, they worked the problem. I was both excited and nervous. I took the papers home to score and could hardly contain my excitement when I saw we would be hitting an all-time high point on the graph! The next day I put on my best sad face and asked the students to come over to the graph. Sounding like Eeyore, I told the children I had checked the papers and had a total class score. Their faces fell. Nick said, "I can tell we did worse." They may have each felt good about their own work, but they had no way of knowing how others had done. I placed the sticky dot 36 points above our last problem, and 8 points above our previous all-time high, which had been many weeks ago. The children burst into a spontaneous cheer that I'm sure must have been heard throughout our whole wing of classrooms. After they settled down, we talked about why we had done so much better. Jen summed it up: "We are learning from what we did before."

I did some reflection. In the first way of doing the problem solving in their groups, children were looking back at the problem, learning from their mistakes. Hopefully, another problem would come along in which they could practice what they learned. But learning in retrospect is not as much fun as having some strategies as one is going into the problem. By practicing the strategies in a relaxed group setting, the fear of the next unknown problem that was to be dropped upon them began to melt away. Roy, who had rarely been able to solve a problem, began to ask if we could do more problem solving. He was one of my last hold-outs on the Math Enthusiasm graph, where he always marked, "I don't like math." Still having trouble and needing plenty of support, he was now enjoying the work. I was not stuffing the learning down Roy's throat. He was eating because he was hungry (see Figure 1.24).

In the following weeks we continued the new plan. I wrote problems similar to ones we had already done. Developing the math stations that let the students experience the strategies needed turned out to be an exhilarating experience for me. Not only did the students have a purpose in completing the tasks; I had a purpose for creating them.

Let's look at the activities and results of some of the next problems. One problem read: Piano Keys. There are 52 white keys on the piano. Kelly decided to try the sound of some of the different white keys. With her right hand, she played 18 keys, and with her left hand she played 10 keys. How many keys were left unplayed? This problem was similar to Bus Seats, and a problem we did after that called, Classroom Seats. For three of the preparation activities, I used egg cartons that had been collected and saved for art projects. I put tagboard wings on eighteen-hole egg cartons to make airplanes. I connected four twelve-hole cartons together end to end with

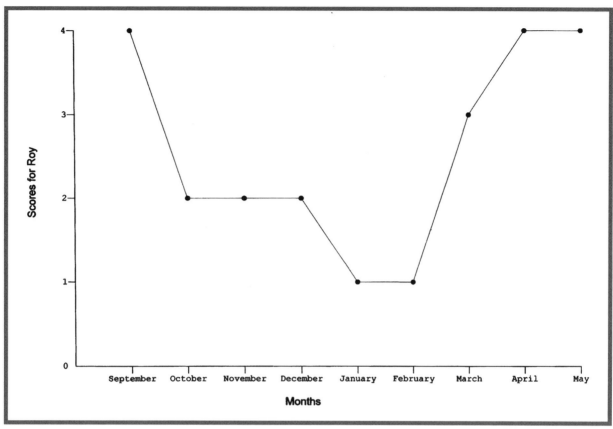

Figure 1.24. Math enthusiasm—student run chart for Roy.

brads to make trains. And I used a thirty-hole flat for the holes to put "Easter eggs" in. These are the problems that accompanied them. (1) Airplane Seats. There are 18 airplane seats. Sitting in those seats are 7 children and 6 adults. How many seats are empty? Children used two different colored rainbow tile for the passengers. (2) Train Car Seats. There are 48 train seats. Sitting in those seats are 17 women and 18 men. How many seats are empty? Linker cubes in two different colors were used for the passengers. (3) Easter Eggs. An Easter bunny decided to dye his eggs with just two colors. He dyed 14 of them red. He let you decide how many of another color to dye. He hid them in little holes in the ground, but he left some holes empty. How many holes did he leave empty? I did not say there were 30 holes in the egg carton flat. I decided finding that number could be part of the problem. Jelly beans were used, and later the children got to eat a few. The last activity in the set used a blank grid with five rows of ten to make fifty squares. (4) Heads and Tails. Samantha began flipping a coin to see how many heads and how many tails she would get. She began filling in a grid that had 50 squares. She wrote H for heads and T for tails. She got 13 heads and 21 tails, but then she got tired and quit. How many squares were left blank? On all of the problems children were asked to use pictures, equations, and writing to show their thoughts on paper. Over a period of four days, each group of five children completed each of the tasks. The following Monday children worked Piano Keys individually. I scored the papers. The sticky dot went up high on the chart, but because it was down two points from our all-time high, the children registered some disappointment. Now that they'd gotten the hang of things, they saw no reason the dot shouldn't just shoot straight up off the chart. So we looked at the results of the other two problems they had done which were similar to this one. We saw the two great frog jumps from the bottom of the chart to our present position. The improvement was obvious! Once aware of this information, the children were pleased.

We had done so miserably on the problem that Peggy McLean observed, I was interested to see how the children would do with division if they practiced strategies first. This would be the test problem: Terica's Money. Terica had 60 cents. She wanted to share it equally with herself and three friends. How much money would each person get? I laid the four practice stations out the day before we were to begin. The children begged all that day. "Can we do the math stations now?" I suppose their enthusiasm wasn't hurt any by the red licorice rope and the marshmallows laying on the back tables.

(1) Red licorice. Mrs. Ayres has a piece of red licorice. Please measure it in inches. If the licorice is divided up among 5 people, how many inches will each person get? At the bottom, there was an additional challenge. Hard question: How many big strings of red licorice will Mrs. Ayres need to give everyone in the class those same number of inches? For this one, children used one-inch color tiles to measure the licorice rope. The tiles were then sorted into five piles. (2) Warren's Porcupines. Warren is designing a new toy to look like a porcupine. He has 4 porcupine bodies (marshmallows) and 33 quills (toothpicks). If every porcupine has the same number of quills, how many will each porcupine get? (3) Ghada's money. Ghada has 85 cents. She wants to share it equally with herself and 2 friends. How much will each of them get? (4) Eating Cheerios. Katie, Amber, Jennifer, Roy, Nick, and Jared want to finish off a box of Cheerios. There are only 72 Cheerios left. How many will each person get if they share equally? Children used paper plates and real Cheerios.

During the week that students worked on these problems, the feeling in the room was that of being at a carnival. "Ladies and gentlemen. Step right up to the licorice problem. Here are some tiles and a tape measure. Try your hand to see what you can do." Children were on task, working independently, or helping one another. Teacher heaven.

But I was nervous the day I gave the students their culminating test problem. We had talked about our activities and why we were doing them. The students knew they were preparing for another test problem. They exuded trust in their ability to succeed, and I didn't want them to feel disappointment when we entered our data onto the wall chart. As students quietly worked the problem, I could see they were more confident than with the previous division problem. Looking like they knew what they were doing, they immediately set about getting coins, hundreds charts, color tiles, and calculators. Even so, I could feel an edge of tension among them. Ideal. When children are given perplexing problems whose solutions require them to reach, reach, reach, and they are playing a game called "See if together we can beat our last score," tension exists. This dissonance brings about their best efforts.

The total class score on Sharing 50 Cents was 37. We came in at 56 on Terica's Money—19 points higher. I would like to have come in 30 points higher, but 19 wasn't bad. The goal is continual improvement. Up, up, up, inch by inch. The children would also have liked 30 points higher, but were pleased with the jump upward.

About this time in the school year we were barraged with a battery of tests—district, standardized, and California Reading and Literature Results Project tests. One day Uzi said to me, "Mrs. Ayres, we haven't been doing any problem solving." The reality was we wouldn't be doing much more. By the time all the test demands had been met, the school year was just about over. After pondering on how to bring a year-end completion to our problem solving, I decided to find out if the children had learned enough strategies to come up with solutions to a problem they had not practiced for. I decided to reintroduce the very first problem we tried to solve, Rides at the Park. It was the only problem of its kind that we had done all year. The way we had been noticing growth was to compare whatever the present score was with one that represented a similar problem we had already done. Students would have no problem seeing a connection between the beginning and the end if both problems were identical. Since they had done so poorly the first time, and so many problems had gone between, I was sure it would seem like a new problem to them. Together we looked at the graph and saw the beginning classroom score of 21. The students eagerly took on the challenge, and I got excited as I walked around the room, watching them with their different ways of thinking. They did remember the problem; they also remembered their inability to do it. This made the final score a victory for them—a whopping 70 (see Figure 1.25)! What a grand way for the children to end their school year, going out winners; going out knowing they had built-in tools for solving future problems. And I was going out a winner, too, with my own built-in tools for *teaching* problem solving. My secret: PDSA.

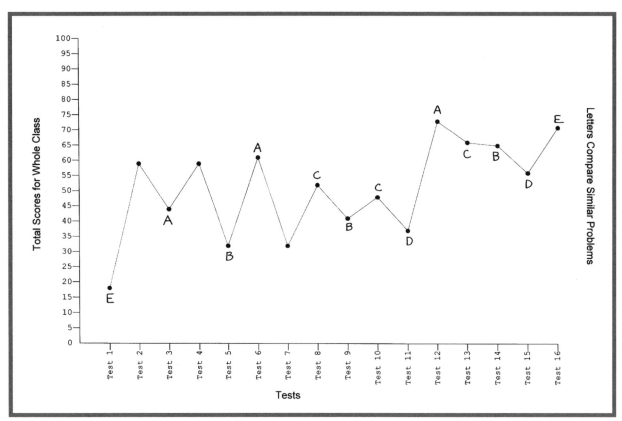

Figure 1.25.   Problem solving—class run chart.

The Histogram

I decided to see what my problem solving results would look like on a histogram. A histogram uses bars to show how often something happens. In my case, each bar stands for a range of correct answers. The number of students with the lowest range of correct answers begins at the left. The height of the bars tells how many students answered that range of problems correctly. When I look at the histogram as a whole, I get to see the forest instead of the trees, except it looks more like apartment buildings in a city than trees in a forest. Some cities have their tallest buildings standing in the middle, some have them to the right or the left. When the tallest bar is in the middle and other bars stair-step down each side, we are reminded of a bell curve, and what we have is a normal distribution. Tall bars to the right or the left invite questions. Using the data already entered in the Class Action program, I created histograms for the first five problems and the last five problems to show the first and last thirds of the year (see Figures 1.26 and 1.27). Five problems with five points each gives the histogram 25 points to break up into five ranges. In Figure 1.26, the tallest bar indicates that the largest group of students got the least number of correct answers. In Figure 1.27 this same range

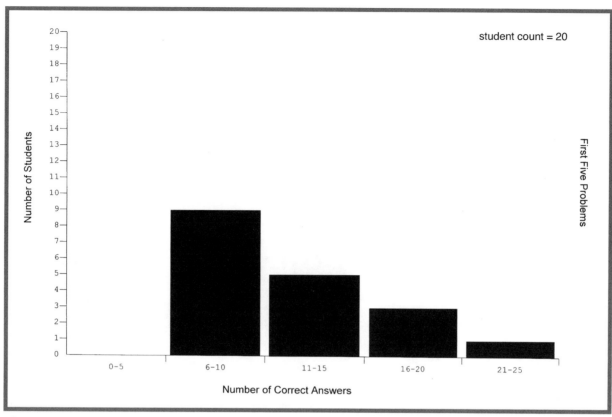

Figure 1.26. Problem solving—histogram, first five problems.

of correct answers has the smallest group of students. I was delighted to see the students had taken stories off of some of the buildings and built new stories on others. Now less students lived in apartments 6–10 and many more lived in apartments 21–25—the better side of town.

About this time, I was digging through some of the hand-outs I received when I participated in the CSU California Mathematics Project, and I found the following advice for teaching problem solving.

Key Points in Teaching Problem Solving: (1) Students must be MOTI-VATED to engage in problem solving situations. (2) Students must begin by experiencing SUCCESS. (3) Students must FEEL that problem solving is IMPORTANT and feel that the *teacher* believes it to be important. (4) Students must have some TOOLS to work with. (5) Students must have adequate TIME to SOLVE and DISCUSS problems.

It surprised me to find the perfect recipe for conducting classroom problem solving in among papers I had read and put in a binder. I felt a little stupid. Trial and error had taken me somewhere I should have already been. Here were clear thoughts on how to teach problem solving—something I had supposedly already learned. I realized that until I worked through to my own understanding, the problem solving guidelines were just words on a page. I

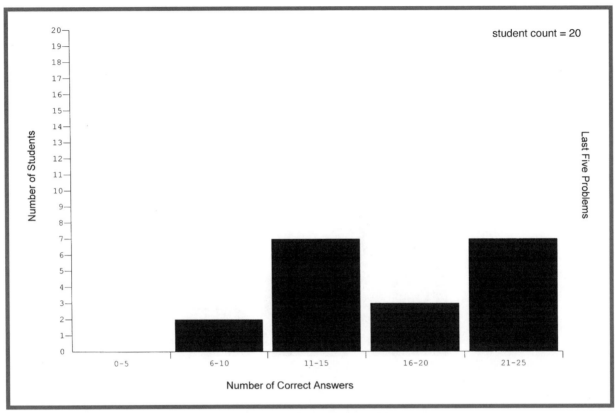

Figure 1.27. Problem solving—histogram, last five problems.

am reminded of the many times I have been exasperated with children's performance, thinking, "How could they be doing so poorly? We've been over all of this." But a more accurate thought would be, "Of course they're doing poorly. We've only read all of this on a printed page." In the Math Project, I participated in doing problem solving myself, but I did not teach it. Now that I have conducted problem solving experiments in my classroom, I can truly say, "I've been over all of this."

Summary

This chapter was written to suggest a plan for creating continuous growth in a mathematics classroom while maintaining enthusiasm. The child and parents see a report card that compares individual progress with a standard, rather than one that uses a curve. A distinction is made between information and knowledge. Information is facts about the past; facts are tools used for thinking mathematically. Knowledge is the ability to use information to create a better future by relating the past to present and future situations through problem solving. A PDSA (plan, do, study, act) approach has the teacher and students constantly checking the results of periodic assessments, in order to monitor growth. Strategies for teaching problem solving are changed as a result of studying the data. Anticipation and enthusiasm mount as the class and teacher watch a large wall graph show a direct correlation between the intention to improve, and the results of their efforts.

Notes

1. David Elkind, *Children and Adolescents* (Oxford University Press: New York, 1974), 51.
2. Lee Jenkins, *Improving Student Learning: Applying Deming's Quality Principles in Classrooms* (ASQ Quality Press: Milwaukee, WI, 1997), 23, 25, 27–28, 29, 31, 37–38, 69, 72–73, 219.
3. Michael DiSilvio, *Class Action* (ASQ Quality Press: Milwaukee, WI). (Class Action can be ordered through ASQ Quality Press, 1-800-248-1946. The Software item number is SW1064.)
4. Marilyn Burns, *Problem-Solving Lessons, Grades 1–6: The Best from 10 Years of Math Solutions Newsletters* (Cuisenaire Company of America, Inc., 1996), 31, 49.
5. Ray Hembree and Harold Marsh, "Problem Solving in Early Childhood: Building Foundations," *Research Ideas for the Classroom: Early Childhood Mathematics* (National Council of Teachers of Mathematics: Reston, VA, 1993), 152, 166.
6. Julie Pier Brodie, et al., *Daily Tune-Ups II* (Creative Publications: Mountain View, CA, 1996), 14, 25, 78, 97, 105, 127.
7. Elaine McClanahan and Carolyn Wicks, *Future Force, Kids That Want To, Can, and Do! A Teacher's Handbook for Using TQM in the Classroom* (PACT Publishing: Chino Hills, CA, 1993), 38–40.
8. Mary Walton, *The Deming Management Method* (The Berkley Publishing Group: New York, 1986), 7.
9. Karen R. Fauss, *Continuous Improvement in the Primary Classroom: Language Arts* (ASQ Quality Press: Milwaukee, WI, 2000), 25–26.

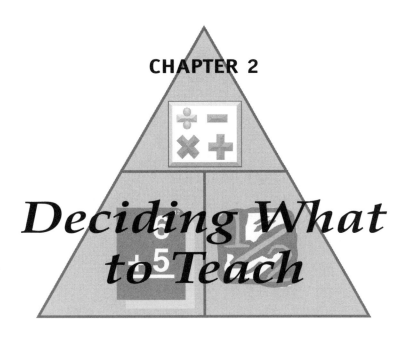

CHAPTER 2

Deciding What to Teach

So much can be shown with a line. A line can create a circle, a mountain, a dinosaur, a car. It can curl around or up and down to write someone's name. The lines on the wall graphs in the classroom were creating uniquely individual drawings of class learning. The problem-solving graph was finally starting to look more like the upward than the downward part of a roller coaster ride. And except for a few butterflies-in-our stomach falls, the mathematical information graph—results of the Enterprise Weekly assessments, (defined in chapter 1 and the glossary) looked like a dipping and swinging ski lift on its way up. It showed me the steady progress the children were making as they continually learned and assessed themselves upwards.

What were they learning? How to shoot free throws, how to follow a recipe, how to do the dance steps. This is the work of collecting puzzle pieces that connect to other puzzle pieces to create the whole mathematical picture. As I pass out the pieces, some children grasp and put them into place right away. Even when the pieces are slipped to them upside down, they quickly maneuver the bubbled edges into position. Others play with them, twist and turn them. Yet others hold the pieces for a short while, and then lose them. I give the same pieces again; I give new ones—more experience, more practice. The children start seeing how the parts fit together to make a whole. Their mathematical picture grows, becomes more beautiful. With literally hundreds of books showing ideas for teaching math information, I must somehow decide which puzzle pieces to pass out so the picture that comes together for the students by the close of school looks similar to the one in my head called End of the Year Goals.

The purpose of this chapter is to look at the pieces that make a math program. When a parent says, "Tell me about your math program," I want to be ready to talk about the different parts, and the significant role each plays. Even if the parent doesn't ask, I myself want to know what those parts are and how they fit together. Until I figure out where I'm going, I can't get there. Although the emphasis in this chapter is on mathematical information, the reader is to understand that in setting up a curriculum plan, mathematical knowledge, the problem solving part, naturally gets planned in as an integral part.

Using Curriculum Standards to Determine What to Teach

With time pressures, choosing what is essential to teach becomes a challenge. Rather than accepting any one program, I use the state curriculum standards as my guide, and then search out the best activities I can find to support achieving those standards. I do not overlook a good math text as a resource for choosing many of these activities. Usually a school or even a district will choose a math text it feels most closely aligns with the state curriculum standards, while looking to see if it also meets its own needs.

As I provide a program that is developmentally appropriate for the age group I'm teaching, I remind myself not to underestimate what children can do. I've had a recent reminder of this with the five-year-old boy my husband and I are adopting. Tanner came to us when he was four. We did not have the advantage of seeing him go step-by-step through each developmental stage of the first four years of his life. From the very beginning, Tanner adored Gary. The two of them began working on projects together, but, developmentally, Tanner was not able to do many of the things Gary thought he should be able to. In one instance, they were building a little cardboard house. Gary was trying to teach Tanner to use a ruler, and convert a small drawing to the size the house would be, using a scale of 1 to 4 inches. I'd had enough teaching experience to know this was a project appropriate for an eight- or nine-year-old child, not a four year old. They each experienced quite a bit of frustration before the project was finished.

With most difficult situations, patience and understanding come with experience and time. Incrementally, I saw Gary lower his expectations, and although they were still too high, as far as I was concerned, the two began working well together. Since then, I have been surprised over and over as Tanner has been challenged with tasks I felt were beyond his developmental ability. After a couple of days, out would pop the understanding. Tanner's brain was obviously working on the problems incognito.

Watching Gary and Tanner together has been a good reminder that I need to continually tantalize students with material beyond their present understanding. Then I must be patient while I wait for understanding to come. One of the definitions of a good problem is one in which the solution is not readily apparent, where uncertainty exists. The brain seeks resolution. It may take a few days, a few weeks, a few months, but once a problem has been presented and the child gets hooked, the brain begins to search for understanding.

At the beginning of each school year, my own brain seeks resolution to a problem: With things to consider like standards, textbooks, and developmental stages, how can I sort out what to teach? Where do I begin? It is comforting to know that I do not need to figure this out all on my own. Rather than working in isolation, the whole process of making curriculum decisions can be done in collaboration with other teachers. Often schools have a curricular matrix, created by administration and teachers, that sequentially moves children from one area of learning to the next, climbing a spiral of steps up through the grades. Ideally, teachers at each grade level and between grade levels get together, and using this document, plan their curriculums in a way they can share ideas and resources. In Deming's words, this would be breaking down barriers between departments, with the purpose of benefiting the whole system.

Using Assessment to Determine What to Teach

With the curriculum standards and the school matrix out in front of me like blinking stars, the children's needs, based upon assessment, determine the immediate lessons to prepare. For me, the Enterprise Weekly, the math information assessment mentioned earlier, has been a useful tool for discovering student needs. (see Figure 2.1). Another tool with a similar format is a packet called Math Capsules[1] (see Figure 2.2). A description reads: "Math Capsules is a management system using grade level pre-tests, post-tests and daily reviews. The grade level pre-tests identify strengths and weaknesses by strand and specific objectives within each strand; the grade level post-tests measure achievement gains or needs for further reteaching; the daily reviews maintain skills periodically to increase long term retention. All questions on the pre-tests, post-tests and daily reviews are keyed to specific objectives to allow constant monitoring of achievement."

With the Math Capsules, students have five problems a day for 50 days. Usually I begin with the 50 from the year before as a review. The teacher can spread them out as desired. As with the Enterprise Weekly, math information is spiraled through the pages; but unlike the Enterprise Weekly, the problems begin simple and move to more complex. I have used them for years now as a morning activity to begin the day.

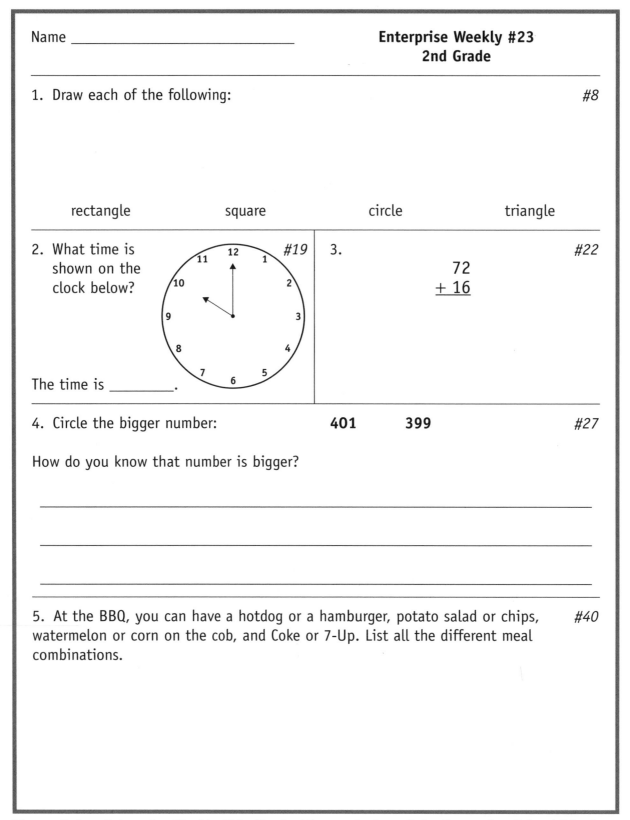

Name _____   **Enterprise Weekly #23**
                                                        2nd Grade
_____

1. Draw each of the following: *#8*

      rectangle   square   circle   triangle
_____

2. What time is   #19   3.   *#22*
   shown on the
   clock below?   72
                                     + 16

The time is _____.
_____

4. Circle the bigger number: **401   399**   *#27*

How do you know that number is bigger?

_____

_____

_____

5. At the BBQ, you can have a hotdog or a hamburger, potato salad or chips,   *#40*
watermelon or corn on the cob, and Coke or 7-Up. List all the different meal
combinations.

Figure 2.1.   Enterprise Weekly #23. Used with permission of the Enterprise School District.

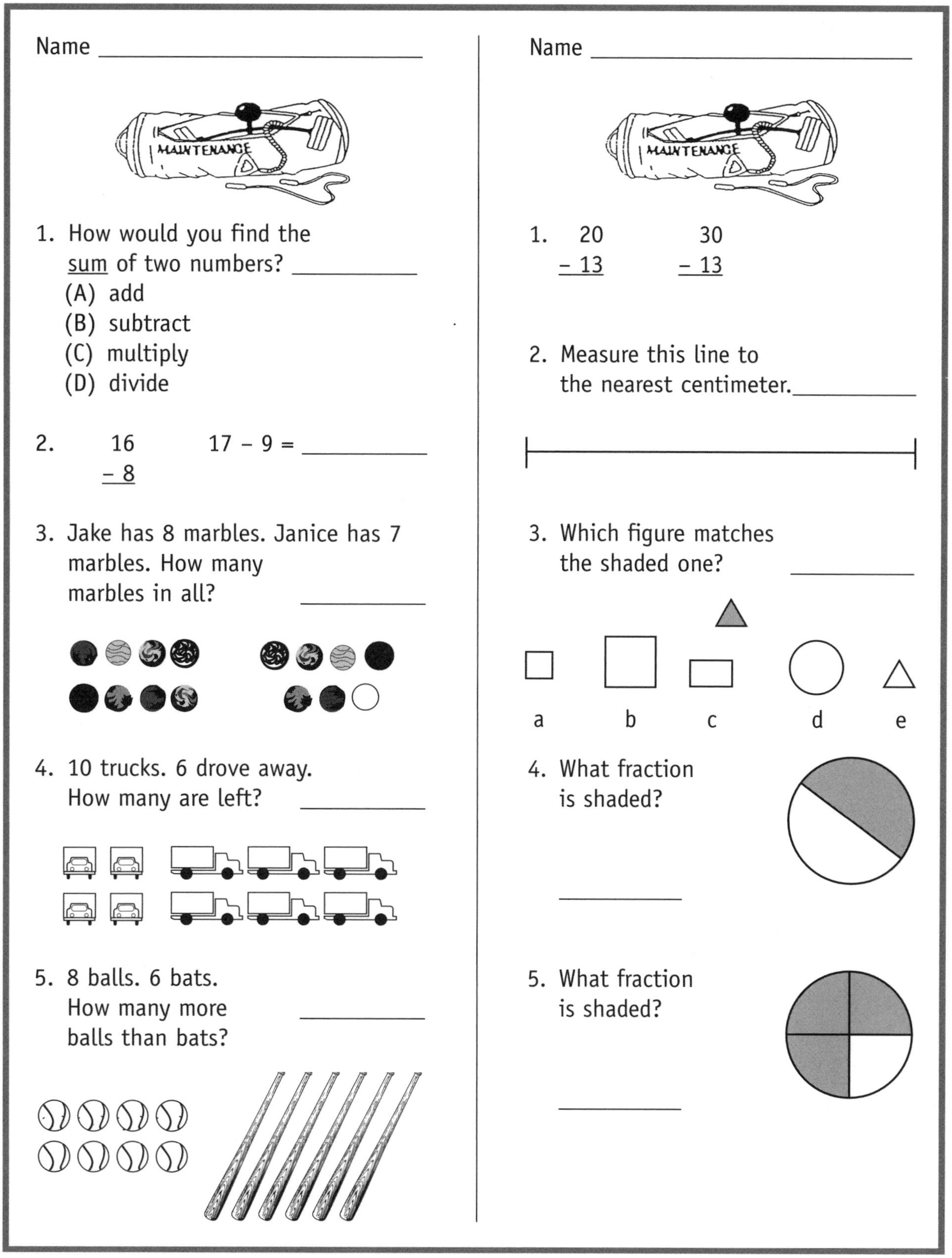

Name _____

1. How would you find the
   <u>sum</u> of two numbers? _____
   (A) add
   (B) subtract
   (C) multiply
   (D) divide

2.   16 $17 - 9 =$ _____
     $- 8$

3. Jake has 8 marbles. Janice has 7
   marbles. How many
   marbles in all?   _____

4. 10 trucks. 6 drove away.
   How many are left?   _____

5. 8 balls. 6 bats.
   How many more   _____
   balls than bats?

Name _____

1.   20   30
     $- 13$ $- 13$

2. Measure this line to
   the nearest centimeter. _____

3. Which figure matches
   the shaded one?   _____

   a   b   c   d   e

4. What fraction
   is shaded?

   _____

5. What fraction
   is shaded?

   _____

Figure 2.2.   Sample *Math Capsules* test. Used with permission. Source: *Math Capsules*
(Math Teachers Press: Minneapolis, MN, 1994).

Using Feedback from the Next Grade Teachers to Determine What to Teach

Another way I decide what to teach is by talking with the teachers in the grade level just above mine. Although looking at the school mathematics matrix gives me similar information, the things the next-grade teachers share are often more specific and personal. I ask, "What information and knowledge would you like children to have when they walk through your doorway in the fall?" During this past year, as I asked, teachers seemed eager to share. I took copious notes, and when they were compiled, I looked to see which things I was supporting and which things needed my attention. It may not surprise the reader to hear that in mathematics, math facts—referring to adding and subtracting the numbers 0–18—came up often. The sentiment was: "If you can teach them their math facts, I can do the rest."

I understand these teachers' frustration. No matter how hard I pound math facts, some children learn them and some don't. As I look at the results of my own class' speed tests, I see a strong correlation between number sense and math fact memorization: Those who lack understanding, do poorly; those who exhibit understanding do better. The feedback I would give teachers of kindergarten and first grade students is this: Please don't have children memorizing facts or algorithms before they have experienced many activities in which they have had opportunities to gain understanding and a sense of number. Now in second grade, my own students have had many of these opportunities. As I continue to do everything possible to build a stronger number sense, I feel students must be studying math facts directly. At this point, the two go hand in hand.

A question about using calculators comes to mind. Maybe having students memorize their math facts is an obsolete learning goal. Can't they just use calculators, which are almost as available as pencils? Many parents would disagree with this point of view, and in fact, often question using calculators at all. The California Mathematics Council has a wonderful publication "They're Counting On Us: A Parent's Guide to Mathematics Education," in which the use of calculators is addressed: "Are calculators a crutch or a tool? Do they hinder or help our children's computational ability? Students need to be able to choose whether a calculator is the right tool for a given calculation. A good candidate for a calculator might be 5,280 divided by 13 (though we insist students know that the answer is bigger than 100 and less than 1000). However, if students continue to use a calculator to do 8×3, they need more work on basic math facts. We are not giving students a choice of tools if they can't do 8×3 in their heads. Calculators and computers are here to stay. We use them in the home and on the job. So will our children. It is more important than ever to understand the mathematics behind real-life situations and to decide whether the number on the display makes sense."[2]

When problem solving in my own classroom, as a support for their thinking, I began by allowing the students to use any tools they wished. Some children did no thinking, but instead punched numbers into calculators to give them answers that had nothing to do with the questions. I decided to ask these students to first show their thinking with a manipulative or on paper before taking a calculator. I encouraged everyone to use the calculator as a way of checking their work, or as a way of completing equations they had not yet learned to solve. Calculators, of course, serve many other purposes. For example, in the problems I chose to use in my classroom this year, I did not include ones in which students look for patterns. A calculator (and a computer) is an invaluable tool for finding patterns— patterns that would otherwise take literally days (weeks, months, years) to discover.

Calculators aside, widespread agreement exists that for facility in doing mathematical thinking, students must know their math facts. Ever since I went to a week-long workshop put on by Math Their Way (a hands-on, manipulative approach to teaching math), I have had a bias against math fact speed tests. Because of this, I have given speed tests only on a limited basis, knowing students would need this experience for their next years in school. I have never tested or challenged my bias until this year.

Wanting to do my best to be a good supplier for the next teachers my students would have, I threw my thoughts about math facts into the brain pot and stirred them around. I wanted to rethink my biases. Dr. Deming said learning comes from testing theories. "Without theory there is no learning, and thus no improvement—only motion."[3] I decided I must test my theory. It was the middle of February. The clock was ticking. Could I find a way to have *all* my students know their addition and subtraction facts, 0–18 by school's end? Probably not. I'd never done it before. From the way the teachers of older students talked, few, if any, had. How about *most?* Could *most* of the children know their math facts by June? Maybe.

Preparation for Learning Math Facts

Before I tell you what came out of my thinking, I would like to share some of the activities I did up to this point to give students experience, understanding, and a sense of number. In this list of activities will be a few games from *Skateboard Practice: Addition, Subtraction* by Elsie Robertson, Mary Laycock, and Peggy McLean,[4] but the book itself belongs in a teacher's library. I use almost every page in the book to teach addition and subtraction.

(*Skateboard Practice: Multiplication, Division:* by Peggy McLean and Mary Laycock[5] is an invaluable resource for teachers of grades 3–6.) Games that I refer to can be found in the appendix.

1. In January of first grade I assigned each student a number ranging from 5–12, which means three or four students each had the same number. Students with less understanding were assigned the lower numbers. When I excused students to recess, I said things like: "Three plus two is excused; five plus six is excused." Students needed to figure others' sums as a way of determining whether or not it was their own. In second grade, I again assigned numbers, this time from 12–16. Subtraction equations were also used: "Twenty minus eight is excused; eighteen minus six is excused."

2. From *Skateboard Practice: Addition, Subtraction,*[4] the children learned to play Ten Spots, a domino game that has them making combinations that add up to ten.

3. Students learned to play Shooting for 10's, also another game from *Skateboard Practice: Addition, Subtraction,*[4] This is another game where students make combinations that add up to ten.

4. I taught them a song called "The Doubles" which we sang often.

5. I periodically dressed up as Señora Ayresa, a fictional character who snuck into the classroom when Mrs. Ayres was not looking. (Señora Ayresa always seemed to know when Mrs. Ayres had stepped out of the room for a few minutes.) Señora Ayresa was dramatic and playful. She taught the students how to check subtraction by adding, and how to check addition by subtracting. Incidentally, the children *loved* Señora Ayresa. They kept asking, "When is Señora Ayresa coming back?" I would say, "Oh, she's on a trip to Mexico right now." It was my way of putting them off, but they wouldn't let up. "When is Señora Ayresa coming back?" What choice did I have? To appease the children, I had to keep asking Señora Ayresa to cut her vacations short.

6. My husband came up with the idea of adapting the card game of blackjack so students could practice math facts. He called the game Closest to Twelve. Students who were more capable, played Closest to Twenty-one. The rules are similar to blackjack except there is no gambling. The children loved the game.

7. I adapted the use of the Circle in a Circle structure I learned in a SDAIE (Specially Designed Academic Instruction in English) class. Students stood in a double circle facing a partner. Children on the inside circle each held a pile of math facts that were all the same. One student might have been holding a stack of eight plus eight. Another, nine plus six. Upon a signal, each student on the inside circle showed

his/her math fact to the partner. The partner had to give the answer in three seconds or the card became his/hers. The outside circle then moved one person to the left, and the process repeated itself. When everyone on the outside returned to his/her original position, partners switched places so the outside circle became the inside circle. Whatever flash cards each student had when the game was over was his/hers to keep at school or to take home for study.

8. Using an idea I got from a Mary Laycock workshop, I created Walk-Around Math Activities. (see Figure 2.3). On pieces of 12" × 18" construction paper I wrote math information questions: "What time does this clock say?" "What time will it be in 30 minutes?" "How much money is this?" "Do you have enough to buy this toy?" "What number is this?" (Students look at a grouping of base ten blocks.) "How long is this line in inches?" "In centimeters?" With the Walk-Around Math Activities, I always included addition–subtraction exercises, such as "Using these two groupings of buttons, write four related math equations." The questions on the pieces of construction paper were numbered and placed around the room. Starting anywhere they wanted, students used their own numbered papers to write answers, as they went from question to question. If used as an assessment, they worked independently with little or no talking. If it was a cooperative learning lesson, students talked and helped one another. The walk-around math activities were not just an enjoyable hands-on experience for students, they were also a good way to do a periodic review of math information. I've done it many times, and never has there been a time I have not had enthusiastic participation by *all* of the children, regardless of ability. I usually set aside three or four challenge questions for students who worked through the main body of questions quickly and accurately.

9. Students used Workjob boxes to create equations. These are materials I made myself using the suggestions from a book called *Workjobs II*.[6] As a way of clarifying, I will describe one of the boxes. The box that says Watermelon has eight 4" × 6" cards, each with a slice of watermelon made of felt glued onto the card. These sit in half of the box. In the other half sit the watermelon seeds—navy beans spray-painted black on one side. A child working with this box would lay out the eight watermelon cards, set seeds on each watermelon, some black and some white (see Figure 2.4), and then place appropriate equations under each: Four white beans and six black beans might be the equation 4 + 6 = 10. The Workjob boxes have been a valuable number resource for many kinds of math activities. Later in the year they were used for multiplication and division equations.

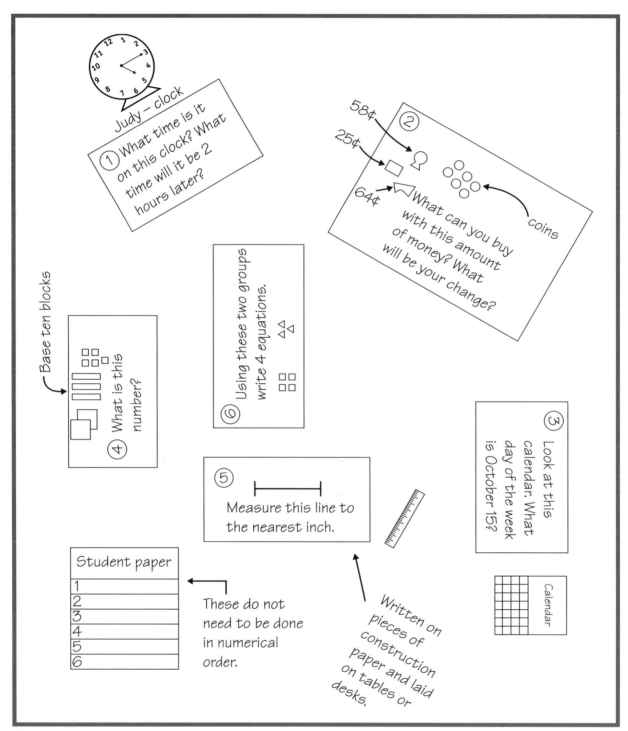

Figure 2.3.   Sample walk-around math assessment.

WATERMELON   **Activity**

The child sets out various quantities or creates problems by placing seeds on the slices of watermelon.

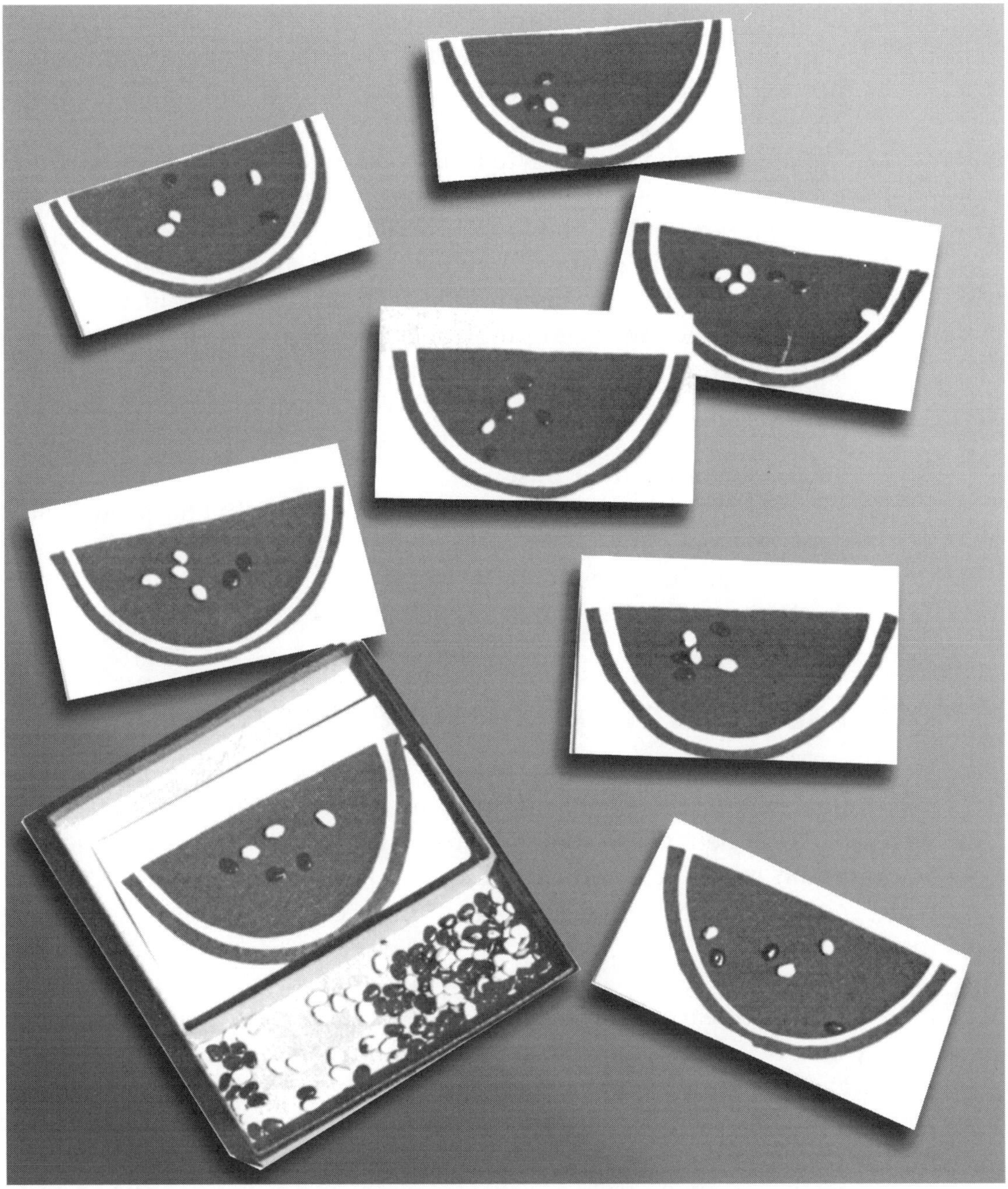

Figure 2.4. Workjob cards. Used with permission. Source: Mary Baratta-Lorton, *Workjobs II: Number Activities for Early Childhood* (Addison-Wesley Publishing Company: Menlo Park, CA, 1979), 58.

10. From *Skateboard Practice: Addition, Subtraction*,[4] I used Hungry Bug Addition, a game using ten as a way to learn addition facts with answers greater than ten. Yet another game from *Skateboard Practice: Addition, Subtraction*,[4] Spin a Flat, has children doing lots of number combinations while also teaching place value.

11. Quizmo (Addition, Subtraction) was a game children played when they had a substitute. This was played like Bingo, but the students needed to add or subtract, then place markers on the sums or differences. Quizmo can be found in math materials catalogs.

12. We used Cuisenaire Rods to build understanding of addition and subtraction equations. Cuisenaire Rods, as well as books suggesting ways to use them, can also be found in math materials catalogs.

13. In place of having the children get bogged down in pages of computation practice, I began giving them "Three Problems" just before lunch each day. They received problems according to their understanding— some with regrouping, some without. I could glance quickly to see if the problems were done correctly. With lunch impending, the children worked quickly as well as carefully, since those who made errors stayed to correct their mistakes.

14. I gave a district-created speed test once a month and plotted student growth using the Class Action computer program. This was a 55-problem addition test with math facts from 1–18. Students needed to get 50 correct in 5 minutes to pass the test.

The Math Fact Learning Experiment

We did these and many other activities. With all of this experience, some of my students were able to commit to memory a number of math facts; others were not. What better time than now to delve into my bias about speed tests? My question could be, "Which will bring about the greatest improvement: math speed tests, or partner-study with flash cards?" The students could be my test subjects.

After explaining to the children the dilemma of not knowing which was the best way to learn math facts, I allowed each to choose by secret ballot, which way he/she would like to study. Fourteen children chose flash card practice; six chose speed tests. As one might expect, the six who chose the speed tests do well on speed tests. Although the two groups would practice in two different ways, as a baseline for noticing improvement, I needed to give a beginning speed test to the entire class, and then continue to give it in periodic intervals. To get my beginning base number for each group, I created four tests of 55 facts each. Two were for addition and subtraction facts 0–10; the other two for addition and subtraction facts 11–18. Children needed to get 50 out of 55 correct in 5 minutes to show mastery of each test. All children began with addition facts to ten. Those who passed went on to the next test. The last two children to keep going did not pass the final test—subtraction facts 11–18. The total number correct for both the speed test group and the flash card group gave me the base numbers from which to begin. The two groups would practice, each in its own way, either with a speed test, or by studying with flash cards, until it was again time to have the whole class take another speed test to assess growth.

Two comparative run charts were put up in the room calibrated to the number of students in each group—calibrated, meaning that as the number of problems correct were charted, each dot went up proportionate to the number of students in that group, and so could be compared. Above this chart was the question, "Which is the best way to learn math facts—flash cards or speed tests?" Beside this graph was another. The words above it read, "We are all learning." On this graph I combined all the results to show whole class growth.

As you see, this was a game, not a controlled test. Most of my more capable students chose the speed tests and were raring to go each day as they posed their pencils, ready to be timed. Many of those studying flash cards, alone or with a friend, were not as focused; some still had not yet memorized their adding facts one to ten because their sense of number was weak, so this was hard work. I expected the speed test group to do better, and they did. But after a couple of weeks, all students became more motivated to give their best effort. This was probably because after I scored the whole class speed tests, I told each child how much he/she had improved personally since the last test; and the graphs on the wall brought into the spotlight the reality of what their efforts could accomplish. A wonderful surprise came when I gave my last district-created speed test toward the end of the year. All but four students passed (see Figure 2.5).

Math Strands

Mathematics is broken into strands. By understanding and teaching these strands, teachers braid a strong rope. Seven strands were in place during many of my first years of teaching. Recently, seven was changed to eight, and now the National Council of Teachers of Mathematics has broken the body of mathematics into ten strands: (1) Number and Operations; (2) Algebra; (3) Geometry; (4) Measurement; (5) Data Analysis and Probability; (6) Problem Solving; (7) Reasoning and Proof; (8) Communication; (9) Connections; (10) Representation. Considering what is developmentally appropriate, all strands are taught K-12. Wow! Algebra taught at the kindergarten, first, and second grade levels? This is a radically different program than what I received as an elementary student. Math facts and arithmetic, which made up most of my elementary math education, are in the " Number and Operations" strand. Parents who have not

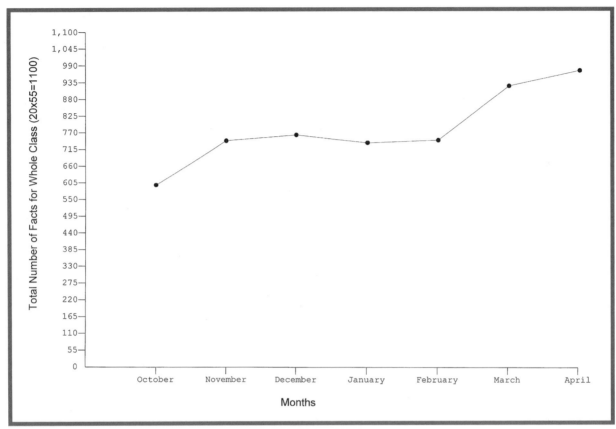

Figure 2.5.   Math speed tests—class run chart.

been educated as to what mathematics encompasses, or who went to school when I did, often think arithmetic *is* mathematics, rather than a part of what is now seen as ten strands of knowledge. Because the number strand is what parents are most familiar with, I use it as my umbrella. It puts parents at ease, and because it permeates all other strands, I can use it as a vehicle to educate and bring them, as well as the children, into the beauty of the whole picture that is mathematics. A rough formula that works for me is is having half of the concepts come from Number and Operations, and the other come from all the other strands. I develop Number and Operations first, and then it becomes a vehicle for moving through the other strands. However, even from the beginning, I am weaving parts of other strands into the Number strand. If I were weaving a piece of cloth in which I wanted many beautiful colors and textures, I would not weave half of the cloth in two colors and then go back and try to slide the other yarns in. Weaving colors concurrently creates a more beautiful and complete tapestry. As I look at the big picture, I can easily include a nice, bright, yellow stripe of geometry somewhere in the middle, and a strong blue stripe of measurement further down the cloth.

In creating a scope and sequence for my year, I remember to begin, at a manipulative level, those concepts that the students will be doing with pencil and paper two grades later. What a boon for the third and fourth grade teachers if I have been doing multiplication and division at a manipulative level with my second grade students!

For beginning and ongoing support in creating a mathematics program, endless possibilities can be gleaned from math texts, lesson idea books, math workshops, other teachers, and one's own creative thoughts. To keep all these lessons from being an organizational nightmare, I use a math file cabinet. As I finish a successful unit, I put the directions and student samples into a folder for the next year. Using some of the strand names as headings in the file cabinet is a useful way to begin.

My own understanding of the distinct characteristics of the math strands was greatly enhanced upon reading a book called *Mathematics: Model Curriculum Guide, Kindergarten through Grade Eight* (1987), put out by the California Department of Education. It was published when teachers were using the seven-strand model, but I recommend it as a helpful guide in setting up a math program. Sample teaching ideas representing each strand are given in the categories K-3, 3-6, and 6-8. Under a section called "Designing the Program" are these subheadings: Alignment of Expectations and Materials, Articulation: What is expected in Each Year, Structuring a Year's Program, Providing for the Range of Students, Homework and the Parents' role, Support for the Teacher Experimentation. "The Program in Operation," covers The Beauty of Mathematics, Mastering Single Digit number Facts, Arithmetic Operations, Reinforcement, Mathematical Thinking, Problem Solving, Concrete Materials, Number Sense, Calculator Use, All Strands for Students, Oral and Written Work, and Assessment.

An even more comprehenshive document is *Principle and Standards for School Mathematics*, the National Council of Teachers of Mathematics document. This book is full of sample lessons that demonstrate teaching in each of the strands at each grade level grouping (Pre-K-2, 3-5, 6-8, 9-12).

Summary

In getting ready to teach math, I would follow the guidelines I have enumerated: become familiar with the state framework for mathematics, look at the district and school math curriculum matrixes, and talk with other teachers. When the children arrived, I would assess to discover their present abilities, then use resources available to create lessons that match those needs that surfaced. I would make some decisions about how to manage groups of students and how to efficiently disseminate materials.

Notes

1. *Math Capsules* (Math Teachers Press: Minneapolis, MN, 1994).
2. "They're Counting On Us: A Parent's Guide to Mathematics Education," (California Mathematics Council: Clayton, CA, 1995).
3. Lee Jenkins, *Improving Student Learning: Applying Deming's Quality Principles in Classrooms* (ASQ Quality Press: Milwaukee, WI, 1997), 24.
4. Elsie Robertson, Mary Laycock, and Peggy McLean, *Skateboard Practice: Addition, Subtraction* (Activity Resources Company, Inc.: Hayward, CA, 1978), 14–15, 26, 27.
5. Peggy McLean and Mary Laycock, *Skateboard Practice: Multiplication, Division* (Activity Resources Company, Inc.: Hayward, CA, 1980).
6. Mary Baratta-Lorton, *Workjobs II: Number Activities for Early Childhood* (Addison-Wesley Publishing Company: Menlo Park, CA, 1979), 58.
7. *Mathematics: Model Curriculum Guide: Kindergarten through Grade Eight* (California Department of Education: Sacramento, CA, 1987), 2–6.

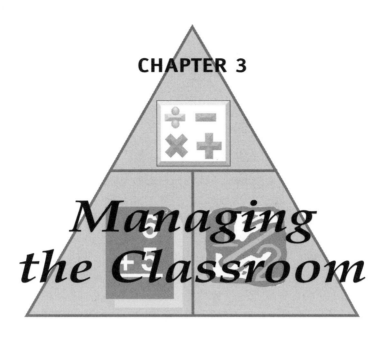

CHAPTER 3

Managing the Classroom

When I first started teaching, I was a great dog and pony show. I loved it. I was the conductor of my classroom orchestra. When I said soft, everything went soft. When I said crescendo, up we went, louder and louder. I read an article that offered the theory that many people go into teaching and law enforcement because they want to be in control. That must have been me. I was in control and it was fun, fun, fun. I just knew I must be a pretty good teacher, too.

It was not until I spent three weeks in the intense training given by the Northern California Writing Project that I viewed myself and my teaching style in a different light. I saw that although what I was doing was tight and classy, it was not the most effective way for my students to learn. Children learn minimally when they are being entertained; when someone else is making the decisions about their learning; when the teacher is doing most of the thinking! Spencer Kagan, cooperative learning educator, said, "Studies of both animals and young children indicate that the more active involvement required of the organism, the greater the likelihood of learning." Clearly, I needed to change what I was doing.

Often, change means things get worse before they get better. I was uneasy realizing the changes I needed to make. To maintain control, I was teaching almost exclusively in a whole group situation. To allow more active involvement, I began breaking students into small groups where they worked independently or with adult supervision. Things got worse. As we experienced chaos and confusion, I found it best to focus upon the beauty of the anticipated improvement in learning, rather than on all the ways things seemed to be falling apart. I looked and saw where improvement was needed and made changes. It took some time, but desire and practice produced solutions. The more confident I became, the more I included students in making decisions about their day and how to make things better. Instead of the control resting with me, the children became more responsible for their own behavior. I relaxed, the children relaxed, and things began to smooth out. As I look back, I feel this change was a growing up, a maturing into my teaching.

Finding what creates the best results and being willing to change is an essential quality for me as an educator. Do I base my teaching on research that shows the best way for children to learn, or do I clutch fearfully to outmoded teaching methods that I have practiced endlessly; methods in which I make few mistakes, but where there is less opportunity for exciting leaps in student learning?

If one were to peek into my classroom now, he/she would see variety throughout the day. There are mini-lessons with whole group and small groups. I facilitate while others are on stage. Children are encouraged to be responsible for their own learning. I can now tolerate a variety of activities in the room, along with a reasonable busy noise level. Actually, it's more than tolerating. I am thrilled when I look out over the classroom and see children doing things, talking to one another, figuring things out. I can't say everyone is on task 100 percent of the time, but most are engaged in their learning, a different picture than their past peripheral participation, where I stood in front of the class with my entertaining, or not so entertaining talking.

Involve Me, I Understand

Just about the time I begin to forget how hard it is for children to sit and listen to me as I stand in the front of the room, droning on and on with my infinite wisdom, I attend another teacher workshop where I am the student. After sitting for a while, I begin to fidget. I notice how much fun the instructor is having—the one doing most of the learning. As the clock hands snail around some more, I lose whole sections of what is being said. I doodle on my paper, whisper to the person next to me, get up and go to the restroom, and remember what it's like to sit, to be uninvolved.

In one such workshop, I was given the following hand-out:

Ability of Learners to Retain Information
"Studies indicate that, in general, people tend to remember new information in accordance with the following percentages: 10% of what they read; 20% of what they hear; 30% of what they see; 50% of what they see *and* hear; 70% of what they say as they talk or write about something (i.e., accomplish a task); 90% of what they say as they teach the material to a peer."

These figures are only approximate, but they do give an indication why hands-on activities and collaborative groups are successful in the classroom. Students need multiple opportunities to respond, explain, ask questions, and listen. In the words of an ancient Chinese proverb: "Tell me, I forget. Show me, I remember. Involve me, I understand."

To determine what ways to involve students, look to see what will most enhance the child's learning of the concept that is being taught. I will share some favorite ways of managing classroom groups. The reader can make a determination as to which fit his/her learning situations and teaching style.

The kind of classroom organization I choose is dictated by student needs. Whole group lessons are appropriate for introducing new information that few, if any, students are familiar with. When I introduce fractions, I do a number of whole group activities to give everyone language and background for the in-depth learning to follow. In one activity, the children are given different colored strips of paper all the same length. Each child labels the first strip as one whole. The second is folded in two parts with each part getting labeled as one half. This idea continues as the third strip is folded into fourths, the last into eighths. A brad is placed at one end to hold the parts together. The same thing can be done to show thirds, sixths, and twelfths. Or students might use a juice can lid to make circles on a big sheet of paper. With a ruler I demonstrate how to divide each circle into parts. I tell them how many parts to color and an appropriate fraction is written beneath. Students love doing these activities. It gives them a beginning, a point of reference, so that when we are in small groups we have some common understandings to build upon.

A KWL chart is a good whole group activity with which to begin a new unit. Students are asked what they think they already Know about the subject, what they Want to find out, and then when the unit is finished, the chart is brought out again so they can talk about what they Learned. Asking what students think they already Know about a subject can be eye opening and sometimes amusing.

Once a new unit has been introduced, one of the most effective things I can do is discover where each child is now in understanding, and then move him/her from that point forward. Finding a child's present understanding can be accomplished by using assessments based upon the standards I wish to achieve. As I continue to give periodic assessments, I can discover if what I am doing is working, by using visual pictures provided by graphs.

It is easy to reason why small groups work. Rather than shotgun teaching where I pepper everyone with general mathematical information, I use assessments to move children in small groups from where they are to their next realization. These groupings allow for personal attention. And because it is personal, the children are more easily excited and motivated, more likely to stay on task. Whole group teaching takes less planning and preparation time, but the trade-off is a slowdown of student learning. I am now always asking, "How can I get more small group time to work with students?"

Roy Colby Sarah David Tammy	Tara Susan Mike Billy Amber	Megan Carrie Jerry Chan Roberto	Manuel Ricky Karen Leslie Robin
Race for a flat	Counting jars: estimate and count	Base ten "Build and Count"	Handful of peanuts
Cuisenaire rod equations	Coin stamps	Workjob equations	Junk box sorting
Mrs. Ayres	Mrs. Ayres	Mrs. Ayres	Mrs. Ayres

Figure 3.1.   Work stations.

Work Stations

This year I chose a model set up in a book called *Guided Reading* by Irene C. Fountas and Gay Su Pinnell.[1] It is a system developed for language stations, but can easily be converted to math stations. In the front of the room, I draw a grid on the white board. It has 12 squares. Students' names are placed on movable cards in four groups, each set appearing above one of the vertical columns (see Figure 3.1). This represents four days of work stations, each group doing all the tasks in a vertical column one day, and then moving to an adjacent column the next day. Labeled activities with written directions are placed around the room. I go over the directions for each task before we begin.

Often a theme will run horizontally across the work board, so that each day a child works on an activity that relates to that theme. For example, if the theme was fractions, a group of students might be working with fractions in the cooking area, the next day move to fractions with pattern blocks, the third day do a paper folding activity teaching fractions, and the last day play a fraction game Race for a Pancake.[2] The theme could also run up and down, meaning the students in that group would do the fraction activities all in one day. I tend to spread theme activities across the week, so that if a child is absent one or two days, he/she doesn't completely miss an in-depth area of study. Also, moving from one aspect of math to another in one day requires the brain to switch gears, make distinctions. There is also the aspect of letting an idea sink in for a night, coming back the next day, and having added insight, as a similar activity is tried. An ongoing theme that occupies one of my horizontal columns each week is regrouping games. The book *Place Value and Regrouping*

Race for a Waffle #1

EQUIPMENT: 21 whole waffles and six waffles cut in ½'s, ¼'s, ⅕'s and ¹⁄₁₀'s. Dice labeled ½, ¼, ⅕, ¹⁄₁₀, ¹⁄₁₀, ¹⁄₁₀.

NUMBER OF
PLAYERS: Three. Two players have the race and one player is the banker.

RULES: Players throw die and receive corresponding parts of waffle from banker. Players trade ten ¹⁄₁₀'s, five ⅕'s, four ¼'s and two ½'s for whole waffles.

NOTE: Game may be played with die labeled .10, .25, .20, .50, 1.00, .10 or with die labeled ½, ¼, ¾, ⅖, ³⁄₁₀, ⁷⁄₁₀.

WINNER: The player with the most waffles.

WAFFLE PATTERNS

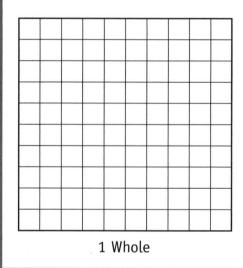

1 Whole

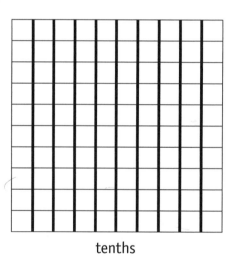

tenths

Figure 3.2. Race for a waffle #1. Source: Lee Jenkins & Marion Nordberg, *Place Value and Regrouping Games.* © 1977, Activity Resources Co. Inc. Hayward, CA. Used with permission.

Games by Lee Jenkins and Marion Nordberg[2] gives page upon page of regrouping games. A partial perusal of the table of contents will give the idea of the variety that is possible: Playing Games in Reverse, Race for a Bean Raft, Race for a 1000 Block, Race to Cover a Trapezoid or Hexagon, Race for Pancakes, Race for a Waffle, Race for Real Money, Millionaire Race, Race for Metres, Race for a Day, Race for a Week, Race for a Year, Race for a Century, Race for Happiness. Taking these ideas, the teacher can change the numbers on the dice to create appropriate challenges for the age group and ability of the children playing the games. For example, in the game, Race for a Waffle (see Figures 3.2 and 3.3), one group may have dice in which the children match

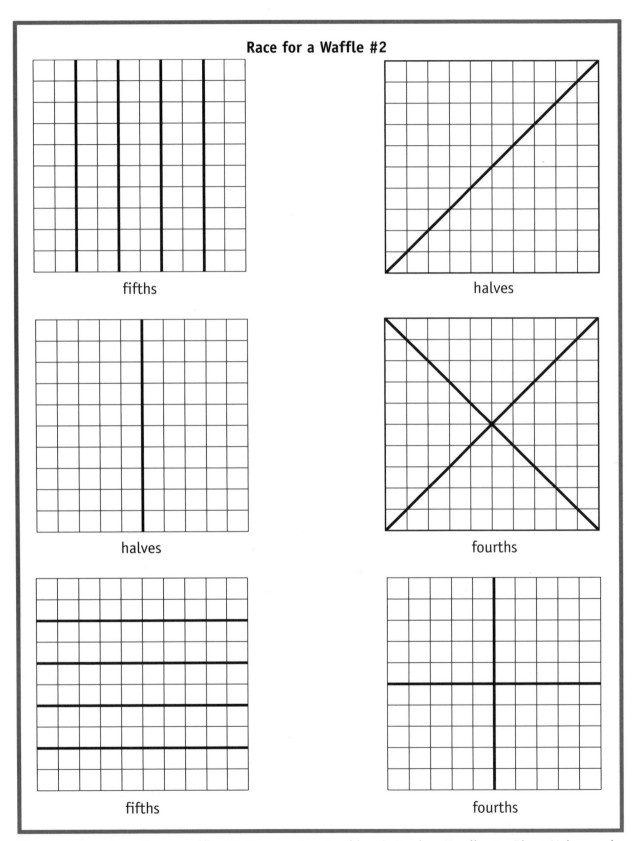

Figure 3.3.   Race for a waffle #2. Source: Lee Jenkins & Marion Nordberg, *Place Value and Regrouping Games.* © 1977, Activity Resources Co. Inc. Hayward, CA. Used with permission.

fraction pieces like ½, ¼, and ⅓ to the side of the dice they roll. The next day another group plays the same game, but their dice have different fractions on the faces: 2 ⅔, ¾, 1 ¼, ⅔, 1 ⅓, 1 ½. It is a good idea to keep a supply of blank dice on hand for this.

Another horizontal column has my name in each square. In an hour's time I am able to work with all four groups, while the other students are involved in their activities. This also means I get to work with each group four times in the week. Children in the rest of the room do not particularly stay together. They know they may work at their tasks independently as well as with others in their group.

The work stations have an extra bonus: The students *love* them. And my, oh my! So would I! They don't have to stay in their seats; they get to choose which activities to do first; they can talk to each other as they work. And most written work is related to manipulative activities. Fun, fun, fun! I am strategically placed so that as I work with my small group, I can watch the other children, and pull the wayward ones back on task in case they thought it would be even more fun to roam and frolic.

Collecting and Checking Papers

Using the format of the work stations, I experimented with finding a good way to collect the papers that were required of many of the manipulative stations. At first I was having children put their completed work in the Finished Box, incompleted in the Unfinished Box. This gave me a hodgepodge of papers that took too long to sort. I now have a box with a hanging file folder for each child. As a paper is finished, the child files it into his/her own folder. Unfinished papers can also be put in the folder, or into the Unfinished Box to be done during a finish-up period. This way I have a portfolio of each child's completed work. If a similar format were used in first grade, each child's folder could be laid upon his/her desk at the beginning of the period—an easier target to hit than a folder in a file box with many others.

Whether or not all work must be completed is dependent upon time constraints and a child's pace. Students know that if they get very involved in one of the stations and wish to spend more time, they are not required to finish everything. (However, if a child is using his/her time unwisely, I may insist that every station *must* be completed, on his/her own time, if necessary.) Too many times in the past I have made children leave an activity where they were just getting the hang of something, and their interest was piqued. My attitude was, "Hurry, hurry, children. I know you didn't get enough time to finish, but I have more things planned for you. The time is passing. Hurry, hurry."

Sometimes it is appropriate for students to finish everything. Careful looking is required to make a distinction. When do I let a child stay involved, and when do I have him/her do the minimum required and move on? I have had a difficult time giving up this business of having everyone finish everything. I spent many years of teaching requiring every child to finish every piece of paper. Control, control, control. Change came as I began to ask myself: "What is the trade-off? Does the virtue of finishing everything outweigh the advantage of uninterrupted involvement in a learning activity? Am I filling my needs or those of the students?" Tough questions!

As I mull these thoughts, I see that the perfect question to ask is, "In any given lesson or circumstance, what is best for student learning?" What I do must vary with the activity, and the uniqueness of each child. I have realized there is not so much a right way as there is a need for constant questioning, mulling, and adjusting.

The work stations are most effective if students can get feedback on what they have done. Is receiving a stack of their own checked papers the feedback they need? It *is* a form of feedback, but getting all the work station papers checked has seemed an overwhelming task. And my experience is that second-grade students usually don't pay much attention to the marks I put on their papers. It is a lot of time spent for a marginal benefit. During my control years I checked and marked *all* the papers, which left me less time for analyzing student work and conducting assessments, meaning less time to plan for children's needs. Asking myself again about the trade-off, I have compromised my terms. I now peruse each folder to pull papers that show the student is having difficulty. These go into my Students Who Need Help folder. When parents or older students come into the classroom as volunteers, I hand them this folder so they can work with students individually. Or I find a time to work with the child myself, either individually, or in a small group, in which there are other children having problems with the same concept. If enough children are having a problem with the same thing, I do a whole-class lesson, or a review in my small groups.

After I have pulled the papers of students having difficulty, I still have all of the others. Often, papers that accompany a manipulative station are difficult for parents to check, but sometimes I have volunteers go through the folders to mark the ones they can. Once in a while, after pulling selected pages, I toss the rest or send them home with no corrections.

What?! Toss them? I have a good analogy for this: A friend of mine removed the fruit trees from her property, replacing them with fruitless ones. They flower like real fruit trees, but they bear no fruit. My friend had other things to do with her summers than attend to the fruit on her trees. And she couldn't live with the guilt she felt when the fruit dropped to the ground, wasted.

What would be an alternative? I say, have fruit trees—lots of them. Eat some fresh fruit, make jelly if you want, or a fresh apple pie. Share with others. Let the rest rot on the ground and don't feel guilty. I suppose it is best to use all the fruit and check all the papers, but when I can't squeeze any more hours into a day, I toss papers and don't feel guilty. This is done with the knowledge that I have looked through them to discover students who need reteaching; and with the knowledge that students learned in the process of doing the work required to complete them.

Using a Flowchart

Interruptions are another management concern. To allow me to focus with my small group, sometimes I begin the work period by saying, "Ask three before you ask me." The point is driven home when I ask the next child who interrupts to give me the names of the three people he/she asked for help. This nudges the child into remembering what to do the next time he/she needs help.

I discovered an even better way to get children to ask others for assistance— the flowchart. "Flowcharting is a way one can get a snapshot of each process within a system."[3] Not only did it improve the station time immensely, but it delighted the children. On past occasions I had seen flow charts, but I had never gotten very excited about them. To me, the flow chart looked like an overworked road map that, if a person wanted to take the time to figure out, would end up stating the obvious; a mind game with very little practical use. Then I read *Tools and Techniques to Inspire Classroom Learning* by Barbara A. Cleary and Sally J. Duncan.[4] Chapter by chapter I tried the various techniques to see what worth any of them might have for me. When I came to the chapter on flowcharts, I thought, "What the heck? I'll give it a try." I made a big flow-chart for our work stations, showing all the places a child should try to get help before coming to me. Interestingly, it took me awhile to figure out how to make it! Looking at a completed flowchart laid out neatly on the page of a book is quite a bit easier than making my own.

I finally succeeded in getting all the boxes, lines, and arrows on one big sheet of paper. I posted it in front of the room at the beginning of our work station time. For two or three days, we began our work period by tracing the different arrows that spelled out each place a child should go for help before coming to me (see Figure 3.4). Even though I had told the children the same things in rather general terms, the chart clarified and made each act specific. It was something visual they could latch onto. If they couldn't remember, there it was, a ready reference. Truthfully, the real reason it worked was because the children saw it as a game, a treasure hunt, a maze. They used their fingers to follow the trails, and interruptions dwindled to almost zero.

After some time had gone by, I decided the chart was no longer needed. I tried to leave it in its place in the drawer, but the children would not allow it.

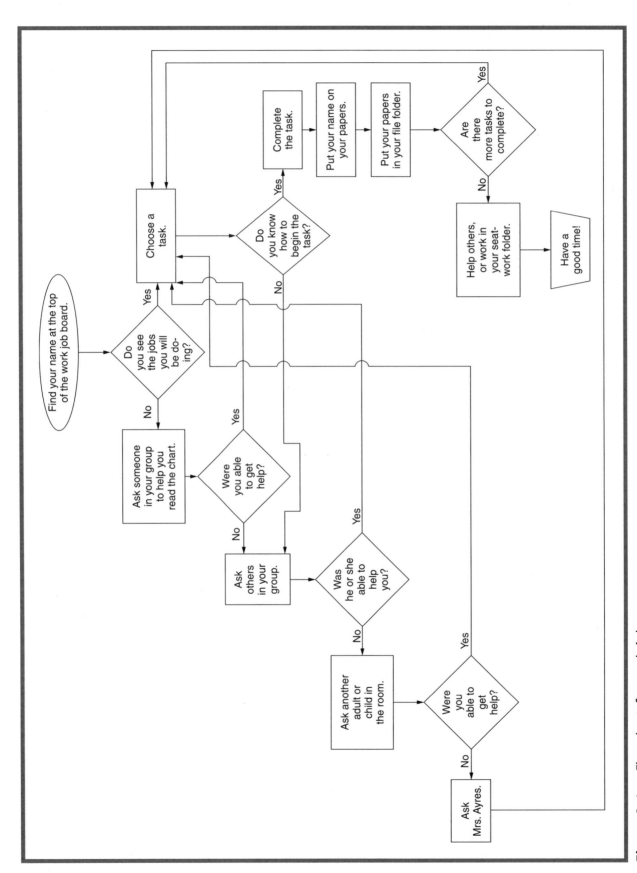

Figure 3.4. Flowchart for work jobs.

They wanted their flowchart! I was amused and surprised. I was also amused and surprised when some of the children began drawing their own. Now there's something to see!

I don't want to be interrupted when I'm working with a group, but sometimes I need to interrupt the children as they are working in their groups. As I see problems arising, I may want to expand upon directions given earlier; or I may want to tell them how much time they have left. I keep a plethora of little music noisemakers on top of the piano: a harmonica, a recorder, two or three kinds of whistles, a thumb piano, and a bell. Students know if they hear any of these sounds they are to immediately stop talking. We practice until we get it perfect. The children hear the sound; the room goes dead quiet. Besides the noisemakers, there are two words we spell to get the same effect. When I say "L-I-S," the children chorus "T-E-N;" or when I say "Q-U," they say "I-E-T." After I make my announcement, or give my direction, the children go back to their busy buzzing.

Other Station Options

With the *Guided Reading*[1] work station format, there are 8 independent work stations to prepare. (The last four stations are the ones in which I work with each small group.) These stations need to be tasks the students can either do by themselves, or ones other students can help them with. For maximum use of time, they need to be meaningful. It is not a good idea for me to squander valuable student learning time with baby-sitting activities that buy me small group time. But preparing 8 meaningful stations every week is way overwhelming! My solution is planning the work stations for every other week. In the alternate week I have an opportunity to participate in the variety of other ways that mathematics can be taught. We work whole group some days, or we break up into smaller groups to play math games. It is a time I can do the groundwork needed for the next set of stations. This is also a time I can put students into problem solving groups, as mentioned in the previous chapter.

Another thing I might do during the alternate week is to arrange the students into three groups, and rotate them through three activities, all in the same period. With this kind of an arrangement, I can float around the room, guiding and facilitating the students' learning at each station, as needed; or I may have two stations that need little adult supervision while I position myself at a third one, where I can do some intensive small group work. Three rotating groups is an excellent set-up for volunteer parents. They learn what they must do for their one activity, and then they guide all three groups through that particular learning process. This is an arrangement where students are usually required to leave what they are doing to move on to the next activity, finished or not. Depending upon what is desired, work can be finished at another time.

If I am giving students math investigations in which they need to be given more time to solve and discuss, I slow everything down. I use the *Guided Reading*[1] work board approach, with each station having two jobs instead of three. Instead of being with my own group, I go around the room helping individuals or small groups with tasks they are working on. Because there are less work stations to accomplish, some students finish before the time is up. These students are asked to pull out their seatwork folders. I will describe this useful management tool.

Seatwork folder #1, a set of papers I have selected from various math books, has a variety of things to do—measuring, drawing pictures with a ruler, computing, telling time, coloring math puzzles. A student may choose which order to do the pages, and he/she works at his/her own rate of speed. When all the pages are completed, I check them and return ones that need to be corrected. When corrections are made, the child receives seatwork folder #2. My best math students work through the first three or four folders quickly. For this reason, I begin including more challenging work in seatwork folder #4 or #5. Rather than have a very capable student work through the first four, I might just begin him/her with seatwork #5. Quite a few of the students love their seatwork folders, and work in them diligently any spare moment they get. Some prefer to read or write if given a choice of free time, but no one dislikes the folders.

Like many students, I also love the seatwork folders. This is because (1) many students love them; (2) they are something all students can do to fill the last 10 or 15 minutes of a period, or some students can do if they finish their work early; (3) they can be used as an emergency if I have to leave the room; (4) they are a substitute teacher's dream; (5) students get extra practice working with mathematical ideas and computation; and (6) they work equally well for tortoises and hares.

Organizing Supplies

I have talked about different ways of organizing so students can practice mathematical content in small groups, but success with math stations is impossible if supplies and manipulatives are not organized and easily accessible. I have adapted many of my organizing ideas from the Math Their Way workshop I took many years ago. Labeled tubs of manipulatives are placed on shelves that have matching labels. It takes a day before school starts to set this up, but from that time on, students do the rest. During the first week of school they practice using the materials, getting them back into the correct tubs, and into their places on the shelves. As one accumulates more and more materials, additional shelves are required. Sometimes when I have not been able to get all I have needed, I have made my own with cinder blocks and boards.

Students can also set out and put away materials for the math stations each day. This needs to be done if there is not enough room to keep the stations out all week. Students match a geometric symbol hanging at a table or desk in the room with the symbol taped on the side of the tub that has the station materials. One student is assigned to each station tub and places it upon its corresponding table. Presto! Instant set-up and clean-up! On the work job assignment board, rather than writing the names of the activities, corresponding symbols can be used to designate the stations. Even though it seems time consuming, initially organizing the classroom for ease in using math materials and setting up stations saves much time and frustration in the long run. To neglect this aspect is to work harder, not smarter.

In discussing manipulatives, I am speaking from the point of view of a second-grade teacher. However, I do not think children outgrow working with mathematical concepts at the concrete level. I remember how delighted and surprised I was to see the person I consider to be queen of mathematics, Mary Laycock, showing how algebra could be taught with base ten blocks to seventh and eighth grade students. In her class, working with manipulatives, I gained first-time understanding of many math concepts.

I once had a parent complain to me that all we were doing was playing every day during our math period. Her son had said something to that effect. I invited Paula in and kindly explained that although her son thought he was playing, manipulatives were actually sophisticated tools for teaching mathematics. After pulling out the base ten blocks, we worked some problems together. Paula became more and more impressed as we "played." Before she left, I couldn't resist setting up a problem that was difficult to solve without experience with the base ten blocks. She was gracious as she acknowledged the value of seeing math. What I learned from Paula helped me see something, too. I had been using math manipulatives almost exclusively. I realized it would not only set the parents at ease, but would be better for my purposes, if I balanced my program by using more paper computation concurrently with the manipulative work.

I think it would be useful to give the reader a peek at the variety of math tools and manipulatives on the shelves in my classroom. On a level that students can see and reach, one will see Linkercubes, base ten blocks, rainbow tiles, pattern blocks, Cuisenaire rods, tangrams, pentominoes, and geoblocks all in tubs. In a nearby vicinity are empty margarine tubs and other similar small containers. These are used when I want to quickly give a child a small amount of coins, base ten units, or rainbow tiles.

Also available to the children are rulers, measuring tapes, calculators, pattern block and tangram stencils, dominoes, playing cards, puzzles, cardboard and plastic coins, coin stamps and stamp pads, and clock faces with movable hands.

A favorite box is the one that has the label Things to Count. Inside are little bottles, boxes, and baskets filled with acorns, beans, cubes, plastic spiders, coins, toothpicks, marbles, jacks, and clothes pins. To heighten interest, students can be invited to collect their own items for the box. Each container has a label so students may write down the item being counted. On a page with prepared lines and columns students are asked to first estimate, and then count their chosen items. I have used this box often as a math station, or as an assignment for a child having difficulty with his/her math work, with no one available to give the one-on-one attention that is needed. Rather than assigning busywork to keep the child from becoming a discipline problem, he/she can be meaningfully engaged until help arrives. For capable students the box may have little value, but many students at the second-grade level are still needing a better grasp of number. This is yet another way to provide experience, which is the beginning of understanding.

Two other tubs hold the Junk Boxes, a treasure I brought back with me from the Math Their Way workshop. They are 4″ × 8″ × 2″ boxes, each filled with a collection of junk, very much like the things in the counting box containers. The labels read, shells, keys, buttons, golf tees, coins, beads, colored macaroni, nuts and bolts, colored toothpicks, feathers, sticks, agates, bottle caps, and bread tags. This valuable junk can be used for making patterns, creating designs, sorting, counting, and doing number operations. The boxes took awhile to put together, but the return has been a hundred fold.

Student Grouping

The last thing to touch on with math work stations is student grouping. To be most effective with my groupings, I need to ask what serves the students the most for any given unit of work. When I include myself as one of the work stations, I like to group according to student needs. Let's say the theme of the week is fractions, and that pattern blocks are my choice of manipulative. Everyone uses the pattern blocks to learn about fractions, but the lesson varies with each group according to the needs of the students in that group.

When I am moving around the room during the station time working with individuals, rather than working with a group of my own, I like to mix the groups so there are varying abilities in each. This is for two reasons: (1) The more capable students can help those who need help. This is cooperative learning at its best, and I get more free time to work with individuals. (2) Student groupings must be scrambled often to keep children from identifying themselves as being in a low group.

Meeting the Needs of the Struggling and the Bored

Organizing for math work stations is one facet of classroom management. Another is how to meet the needs of children at the extreme ends of the math spectrum—those struggling, and those who are bored because they are not being challenged. As I am plagued with this problem each year, I think about the poster I saw on the wall of Chico State math professor Bill Fisher's classroom. It said, "Teach the students you have, not the ones you wish you had." I committed it to memory.

There are always students with unique needs in each class. I find myself being so busy making plans that move the bulk of the class along, that I go days without helping these children flourish. Of course, if they become behavior problems, I am forced to give them immediate attention. Remembering that motivation to learn is intrinsic, the kind of attention I choose is key. A child's behavior may be telling me, "I'm not relating to what you're teaching. This is not meaningful. I'm disconnected. Help!" Meeting the needs of high and low ability students is very often a time management problem. How can I find enough one-on-one or small group time to keep these children actively involved with learning situations that are meaningful? Here are some of my more successful ventures.

For struggling students I have used cross-age tutors—older students coming into the classroom to work one-on-one or in small groups with children. I have also paid attention to creating an atmosphere where my own students are willing to help each other in kind and effective ways. I have ability grouped so I could work with struggling students during math station time, and I have involved the parents.

Involving parents is a good way for me to leverage my time. A model I continue to use is called The After School Club. (I have also had The Before School Club.) In the class I'm working with this year I chose three students who needed help in both language and math. I sent the following letter home to the parents of these children:

> "I am choosing three children I would like to work with after school. These are not just any three children. They are three children I think really want to learn; three children whose parents I know I can count on. I chose these particular children because, besides their willingness to work, I think they can benefit from extra instruction. We will practice reading, spelling, and math in many ways. Here are the times I would like to set up: Every Tuesday and Thursday, 2:15 P.M. to 3:15 P.M. beginning Tuesday, January 26. If I have a meeting or doctor appointment, I will be sure to inform you at least a day ahead.
>
> "If you can work it out and you want your child to participate, here is what I am asking of you: (1) Once the After School Club begins, it is important to have your child come every time. (2) I want you to commit to spending some

time each week playing the math and phonics games I will be sending home. (3) I would like to have a beginning meeting of the families and children to talk about our purpose and to go over the games. I can come to an evening meeting, if no other arrangements can be made.

"The children in the After School Club will work hard, but I think they will also have a good time. I've created clubs like this before, and usually other children are clamoring to get in."

All parents showed up for the evening meeting. The games we learned to play were The Phonics Game, Yahtzee, Ten Spot, and Race for a Flat. These last two games are ones I use from *Skateboard Practice,* which I mentioned in the previous chapter. In the following weeks, I called once in a while to ask how it was going with the games. The report was always good: They were playing them, and the whole family was enjoying them. Colby's mother told me that his dad had gotten a lot more involved in working with Colby because he knew what to do, and it was fun.

From the very beginning, the three children were led to believe they were smart, and I was going to help them get smarter. Sometimes we worked on things I had not yet introduced in class. I told them I was teaching them first, so they could help others to understand when the time came to do it with the whole class. With this pre-lesson they were ready and excited as I taught the lesson to the class. I was sure to call on them as their hands went up to answer clarifying questions. As before, others wondered how they could get into the After School Club. My three after-school students felt privileged and special. I asked them to not make too big of a deal about being in the club. After all, we didn't want to make the others feel bad.

The club began with the children doing counting strips. This is a place value activity described in *Building Understanding with Base Ten Blocks, (Primary),* by Peggy McLean, Mary Laycock, and Margaret A. Smart.[5] It is another book that belongs in a teacher's library. *Building Understanding with Base Ten Blocks, (Middle)* by the same authors[6] is a book teachers of grades 3–6 will want. Students use a place value board with columns labeled ones, tens, hundreds, and thousands. Adding a unit at a time, children record the number of blocks on the place value board onto their counting strip, which is a four-column grid. Ten units must be exchanged for a long (ten-stick), ten longs must be exchanged for a flat (a hundred block), and ten flats are exchanged for a cube (thousand block). When the children run out of room to record, they tape another counting strip to the bottom. More detailed directions can be found in the appendix. I decided to begin with this activity when all three children showed me they could see nothing wrong with the equation $84 + 88 = 83$. They used shoe boxes to store their materials. As the strips got longer, they were wrapped around empty toilet paper rolls and secured with a rubber band. We put on music and had popcorn. Through the whole winter and spring, the children could not wait for every Tuesday and Thursday afternoon to come so they could attend the After School Club. When all three had completed the long, long counting strip that took them all the way to 1,000, we made a trip to the yogurt shop to celebrate.

Children at the other end of the spectrum, the ones who need an extra challenge, also require special planning. Again, this is a time management problem. This year I have had an arrangement with another second grade teacher in which we take turns gathering both classes together for story time. During the time Patty is reading to my class in her room, I bring a group of children to my room for Math Club. These are children who finish most classroom math easily and are pressing for more problems, more thinking, more math. If for some reason I skip the Math Club, I get nagged endlessly. I am able to use extensions of the things these children learn in Math Club as work to be done during the regular class period, keeping their little minds excited and busy. Two of these Math Club children needed even more stimulation. Amber and Nick's classroom behavior, along with their Enterprise Weekly individual run charts, (see Figures 3.5 and 3.6) let me know it was time to set up a corner where they could work together on challenging material while I was working with others on concepts each of them had already mastered.

The first time I ever used the corner was the year I had Joe in my class. His abilities were so developed, it was difficult to group him with any of the other children. As the class was learning new concepts, I tried to keep him involved by posing questions for him to ponder, but my program was not really set up for someone so advanced. He created a constant discipline problem. As with other problems, I found that once I took the time to look at it directly and began to ask myself what could be done, a solution emerged.

The principal let me take a half a day from my classroom to develop a special program for Joe. Joe's mother was invited to join me on that day. After we got organized, I also had Joe leave class to come help us. Using a collection of science and math books, both subjects of particular interest to Joe, we each began searching for activities that could be done in the classroom. One of the books we looked at was *The Book of Classic Board Games,* collected by Sid Sackson and the editors of Klutz Press.[7] A presenter at a math conference touted it as being exceptional. (I found I agreed, when later, Joe used it in the classroom.) After we had a significant pile, I chose what I thought would be the best tasks, given classroom space and time constraints. From this, a binder of activities was put together for Joe. Our morning ended, and we were all excited.

A special time was set up during each week for me to go over Joe's work with him, first to acknowledge him for what he had accomplished, and then

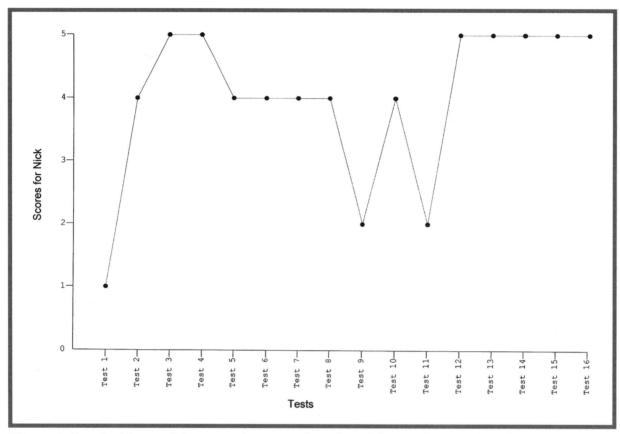

Figure 3.5. Problem solving—student run chart for Nick Carreiro. Used with permission.

to plan what he would be doing next. This was an important part of the arrangement. Joe needed to know that what he was doing mattered; that he wasn't just being stuck off in a corner so I could get him out of my hair. And he did have a corner—a place where he could go during times when he did not need to attend to the learning activities going on in the classroom. I got no argument from any of the children about this. They were in awe of Joe's abilities, and recognized his need for something more. If another child finished what he/she was doing, he/she was welcome to join Joe as he worked on the pentomino and tangram puzzles. Or Joe would teach someone to play a game from the Board Games book. This classroom model turned out to be an excellent way for Joe to relate with the other children. He may have had no trouble with academic work, but he struggled socially.

The model was not just excellent for Joe. Two other teachers became interested in making similar arrangements. Corners were set up in Michelle's fourth- and Beth's fifth-grade classrooms. Each teacher chose students according to her own criteria. Beth had two groups of children she felt would

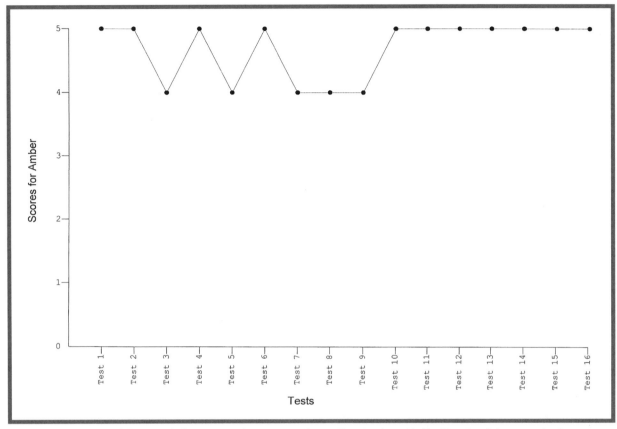

Figure 3.6.   Problem solving—student run chart for Amber Hall. Used with permission.

benefit from working in the corner. After the first group had experienced the activities for a five or six week period, she reassigned it to the second group.

Deciding when students could go to the corner was something that had to be worked out. Beth had her fifth-grade students take turns going during the first 30–45 minutes of the day when morning exercises and whole group activities were taking place. And it was always a place students could go if they were finished with other work. This past year, one of the times I used for students to work on special problem-solving assignments was during our weekly math stations. Each teacher would need to choose times best for his/her own students, in scheduling the times students work at this independent area.

I learned a lot from Beth and Michelle's ways of applying the corner idea. This sharing of ideas from teacher to teacher is an important part of a school system. We leverage our time, and we improve and enhance our teaching by coming together with our problems and solutions. We take a few good thoughts, put them together, get excited, and ultimately create better ones. I like sharing my good thoughts with others—thoughts that were improved upon in the sharing—and I love having others share with me. Following are some great teaching ideas concerning classroom management that were shared with me.

Classroom Economies

I went into the classrooms of both a fifth-grade and a sixth-grade teacher to discover what effective ways they had of organizing their rooms and working with their classes. The day I visited Pam Stephenson's fifth-grade class, I was immediately aware of big bodies in a small room. The first purpose of classroom management is to create an atmosphere where students have the freedom to think and the freedom to learn. Children also need to know they will be safe. The freedoms normally allowed outdoors or at home for growing children must be tightened when so many bodies occupy such a small space. I watched as students took a multiplication math speed test, and then got a whole-class score with calculators for the wall graph. I walked around the room as students worked on individual problem-solving activities. I listened and took notes as Pam gave some whole group instruction. All students were enthusiastically engaged, working, listening, and raising their hands in response to questions. What was Pam's magic? When I got ready to leave, I complimented the class on their wonderful attitudes and said what a pleasure it had been to be with them. Then I said, "Everyone was working so hard and so enthusiastically, for so long! I feel like I've been in a magical classroom all morning. What is the secret?" One student got a twinkle in her eye, reached into her desk, and pulled out chips. She explained that they received chips for participation and hard work. Later they got to buy things with the chips. A few weeks later I asked Pam about the chips. She said students could earn chips for any behavior she wanted to reward: turning in homework, staying on task, cleaning the room, or helping others. They could also have chips taken away to discourage behavior that was disruptive to their own or others' learning. The chips were used as money to buy privileges like going to the restroom during class time, moving one's desk to another place in the room, or getting another pencil. At the end of each trimester, they could use their chips to buy treasured items at a classroom store.

The morning I visited Pam's class, I could see that the students' enthusiasm for learning went beyond the chips. Because Pam was not consumed with the energy it took to deal with discipline problems, she was positive, lively; her lessons dynamic, engaging. Here's the *real* secret: Students want to learn through meaningful activities. They want to notice their own growth; they want to be noticed for what they can do. A good classroom management system frees the teacher to interact with students in a way that allows a nurturing of their abilities.

Ron Johnson is another teacher who uses a classroom economy, this one set up through a computer banking system. Information about this economy can be found on the Internet at http://www.inyopro.com/microecon/. Instead of chips, students receive paper money. A monthly wage is received for showing up at work each day. Using the computer, two classmates bank others' money for them. To spend their money, students write checks or use cash—dollar bills they have designed. Each student has a job he/she is paid to do. The jobs are the busywork that usually munch away at a teacher's day, leaving little time for planning and working with students—things like line monitor, someone to take attendance, teacher assistant, room straighter, computer tech. Students can also receive bonuses for acts of exemplary behavior.

As in the adult world, students pay the expenses of daily living with their hard earned cash. They must pay to use the restroom during class, to buy replacement pencils, or for having lost assignments. They also pay utilities and rent for the use of their desks. Once every three weeks, students spend their money at an auction. At first, Ron was purchasing items for the auction, but now students bring things from home they don't want anymore. Surprisingly, they seem to enjoy this more than the new things Ron was providing. Some students go into debt—another similarity to the adult world. Also, students pay fines when they infringe upon others' rights, turn in sloppy and incomplete work, and break classroom rules.

Ron speaks enthusiastically about this assimilated economy as a management system. The students have ownership in running their classroom lives. Besides being motivational for the them, Ron, like Pam, is freed up to smell the roses; enjoy the uniqueness of each student; notice the subtleties; teach, instead of nag. While students are taking care of the morning housekeeping activities, Ron immediately begins helping students with their daily math starters. Viva la classroom economy!

Any plan I make for going through my school day is useful only if I follow it. One of the ways I get students to support me in accomplishing the daily goals is to put my plans on the board each morning. When I first began doing this, the students showed little interest. Then one day, someone said, "Look! We're playing March Attack for P.E. today!" As the days went along, more and more children started reading through the agenda. They knew that sometimes I changed the plans, but there were those who did not hesitate to let me know if I was skipping something they thought we should be doing. One day, when it was almost time to line up for lunch, Uzi reread the plans, and yelled out, like it was an emergency: "Mrs. Ayres! Mrs. Ayres! We haven't done our three problems!" I responded to the emergency immediately: "Oh, my gosh! Everybody stop what you're doing and come get your problems! We should be lining up right now!" There was an excited scramble, the children worked rapidly, and we weren't all that late to lunch. I thanked Uzi for checking the board.

Teachers who have put their time into planning and using good classroom management techniques make teaching look easy. My twenty-seven-year-old son came and spent the morning in my classroom awhile back. He was with his fiancee, Kristin, who will soon be a teacher. Dave watched as I gave a guided writing lesson, then walked around the room helping children as they wrote. Later he said, "Teaching is going to be so much fun for Kristin. She will be able to get kids going on something they are excited about. Then she can relax, walk around the room, see what they are doing, and help those who are having trouble. I can see it will be a much more peaceful job than the one she is doing now." I said, "Yes, I know Kristin will love teaching," but I smiled to myself. Classroom management techniques are not immediately noticeable to the untrained eye. Dave had no idea the planning that had gone into the peace he was experiencing.

Summary

To summarize, in the mini-system of my classroom, all parts should support the good of the whole. Classroom management is not separate from other aspects of teaching, but done to support the desired outcome. How I organize my classroom is determined by the students' needs, by what must be accomplished, by our goals. Students learn better when they are involved doing and experiencing, as they are guided toward each benchmark. Consequently, discovering ways to have children work in small groups becomes imperative with my classroom management planning. Having materials easily available also supports this intent. The flow chart is a useful tool for creating student independence to allow me to work with small groups of my own. The paperwork students complete is an integral part of the math activities, rather than tacked-on busy work. I look carefully to insure my efforts are having me work smarter, not harder, when checking and returning these papers to the students. By planning to meet the needs of those who are struggling, as well as those who need extra challenge, I am allowed to move forward with the class rather than being constantly distracted with discipline issues. The classroom economy, used in fifth- and sixth-grade classrooms, is another management technique that allows the teacher to move forward without constant distractions. Besides making the classroom run more smoothly, these are management ideas that help students to maintain enthusiasm while continuously improving, which is the real prize.

Notes

1. Irene C. Fountas and Gay Su Pinnell, *Guided Reading* (Heinemann: Portsmouth, NH, 1996), 53–71.
2. Lee Jenkins and Marion Nordberg, *Place Value and Regrouping Games* (Activity Resources Company, Inc.: Hayward, CA, 1977), 3, 13–14.
3. Margaret A. Byrnes, Robert A. Cornesky, and Lawrence W. Byrnes, *The Quality Teacher: Implementing Total Quality Management in the Classroom* (Cornesky & Associates, Inc.: Port Orange, FL, 1992), 239.
4. Barbara A. Cleary and Sally J. Duncan, *Tools and Techniques to Inspire Classroom Learning* (ASQ Quality Press: Milwaukee, WI, 1997), 1–12.
5. Peggy McLean, Mary Laycock, and Margaret A. Smart, *Building Understanding with Base Ten Blocks (Primary)* (Activity Resources Company, Inc.: Hayward, CA), 1990.
6. Mary Laycock, Peggy McLean, and Margaret A. Smart, *Building Understanding with Base Ten Blocks (Middle)* (Activity Resources Company, Inc.: Hayward, CA, 1990), 14–16.
7. Sid Sackson and the editors of Klutz Press, *The Book of Classic Board Games* (Klutz: Palo Alto, CA, 1991).

CHAPTER 4

Forming a Relationship with the Most Important Supplier

A simple formula for Dr. Deming's ideas might be: All parts, working toward improvement of the whole, equals a successful, functional, healthy system. Deming taught that through communication, collaboration, and creative teamwork the parts are not only more able to improve the whole, something magically exponential comes into being. Continuing with the understanding that Deming's philosophy applies to all systems, I began wondering about things that could be thought of as a system. My mind went off on a bee hive. I saw the queen bee as the customer, the one that must be catered to. I amused myself thinking of the children in a school system as the queen bee, all others busily fussing over, and feeding it.

Once a system is identified, its parts can be seen. I like what Dr. Jenkins[1] says:

> Crucial to the understanding of Dr. Deming is the understanding of *system*. Dr. Deming was fond of citing the following example. You could bring to one location the dozen best cars in the world. The top automotive experts in the world could determine which cars had the best engine, steering, brakes, and so on. These best parts could be collected from the best cars by the best experts, and what you would have is a collection of parts, but not one working automobile.

The automobile metaphor is crucial to understanding improvement. To have a working car, all the parts must be in place and they must work together. The seven components of a system are *aim, customers, suppliers, input, process, output,* and *quality measurement.* If any component is missing or not in tandem with the other six parts, one has a collection of pieces, but not a system.

It is clear that no part of a system is more important than another because without any one of them, the system cannot function optimally. Given that, I would like to expand on a part that can be easily passed over in its importance—the supply. Seeing the human body as a system might clarify how supply can affect a healthy whole.

In a body, certain parts must be functioning or the system will no longer exist. Beyond that, how well all the parts of the system work together will determine the well-being. One can limp along miserably for many years, alive, but not healthy. Many things affect the body's well-being, but we know that the quality of the supply—what goes into the body—is paramount to its health. All those little inside factory workers need quality ingredients in order to make healthy tissue, blood, and bones.

Dr. Deming encouraged businesses to form long-term relationships with their suppliers. As they did this, a spirit of goodwill and trust brought about quality products for the customers. Supply is an area of the educational system that is begging for more attention and inclusion. We all know the home environment has a great influence upon a child's success in school. In his book, Dr. Jenkins[1] says, ". . . it is in the best interest of school districts to assign some of their most talented teachers to helping parents and preschool teachers better prepare students for school success."

In talking with colleagues, I often hear frustration about parental support. I teach at a school where the poverty level is above 75 percent, as determined by the free and reduced lunch program. Families are struggling. Can they do more than get their children to school each day, clothed and lice free? Although many lead troubled lives, if they knew how to help their children, and thought it would make a difference, most parents would want to try.

A first step is to help parents see themselves in the role of partner. A prevailing attitude is "It is my job to feed and clothe, give love and support; it is the school's job to educate." Over the years I have tried different things to involve parents as partners, with varying amounts of success. I have sent home newsletters every week to keep parents informed, although I'm not sure they always gets read. Students are asked to read with their parents and get support with homework. I have put on Math Family Nights for parents and their children to come to school and play math games that can be played at home. Some parents have been able to volunteer in the classroom or go on field trips. These are ways many teachers use to involve parents. I discovered another.

Last year I taught first grade for the first time and panicked when I realized I had to teach 20 children to read in just nine months. I looked out the window of my train and saw it just wasn't going to happen. I needed to lay some new track. A fifth-grade teacher agreed to send students after lunch to listen to the children read. I gave a partial sigh of relief. That helped! What else could I do? With this question in mind, the Homework Contract emerged.

The Homework Contract

The beauty of the Homework Contract is that it was individually written for each child. A standard form listed information I wanted students to master including both language and math goals. Using assessments, I either filled in appropriate numbers or highlighted appropriate objectives that were individual to each child. The children had six weeks to master the chosen areas and parents were the overseers.

After sending the contract home, I began by phoning every family to talk with the mother, father, grandmother, or other guardian of the child about it. This was the beginning of establishing rapport with the most important person in every child's life. I explained the contract, and asked if it was something the parent could support his/her child in doing. All said they would do their best. I said I would call back in a week to see how things were going. After that I called on a regular basis. It went something like this: "How is the contract coming along?" The parent might say, "Oh, we haven't gotten it down off the shelf yet, but we will." I often asked, "Can I call again in another week to see how it is going?" "Yes. We will get on it. We'll have a better report for you next time." We were able to talk about a child's lack of motivation. I could answer questions and make suggestions. Often the parents unloaded frustrations they had concerning their own hectic lives. I listened. We became phone friends.

Since this first contract had no input from the students or parents, I invited the parents to look with me over the phone to see if any of the requirements should be changed in order to assure success. I explained that the trick was to create a challenge without having it be so hard the child would get frustrated and want to quit. I also wanted the parents to understand that it was, indeed, to be treated as a contract, which meant "I will work with my child to complete all we have agreed upon." The contract would have been a useless piece of paper without the calls; phoning regularly kept it alive. Through these conversations we built camaraderie and trust. We became partners by creating a common vision of possibility.

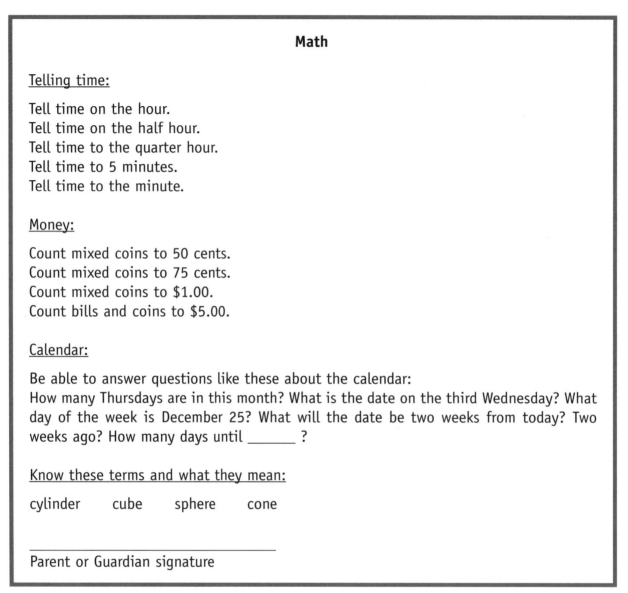

Math

<u>Telling time:</u>

Tell time on the hour.
Tell time on the half hour.
Tell time to the quarter hour.
Tell time to 5 minutes.
Tell time to the minute.

<u>Money:</u>

Count mixed coins to 50 cents.
Count mixed coins to 75 cents.
Count mixed coins to $1.00.
Count bills and coins to $5.00.

<u>Calendar:</u>

Be able to answer questions like these about the calendar:
How many Thursdays are in this month? What is the date on the third Wednesday? What day of the week is December 25? What will the date be two weeks from today? Two weeks ago? How many days until _____ ?

<u>Know these terms and what they mean:</u>

cylinder   cube   sphere   cone

_____
Parent or Guardian signature

Figure 4.1.   Math contract #1.

I tailored the contract as closely as possible to my assessments, which were tailored as closely as possible to the outcome I wished to achieve. With reading and spelling, the contract stated what the child could do right then, as determined from the testing, and what the child would accomplish, with the help of the parent, by the contract deadline. In math I created a generic list of skills—information I wanted students to master (see Figures 4.1 and 4.2). As I mentioned, I highlighted those things I felt each child was individually ready to work on. For example, one child might need to learn to count to 50, while another might be ready to skip count by twos, fives, and tens. Each of these skills was listed, but a child was guided by the highlighting to know

---

Math

I can count by ones to 100.

I can count by twos to 50.

I can count by fives to 100.

I can count by tens to 100.

I can count backwards fluently by ones from 50 to 1 (helps with subtraction).

I can count to 100 by mixing ones, fives, and tens. (This helps in counting money.)
    For example:
    Adult: "Count by tens." Child: "10, 20, 30 . . ."
    Adult: "Count by fives." Child: "35, 40, 45, 50, 55 . . ."
    Adult: "Count by ones." Child: "56, 57, 58, 59, 60 . . ."
    Adult: "Count by tens." Child: "70, 80, 90, 100 . . ."

I can count by threes and fours to 50 (or as close as you can get to 50).

I know adding math facts with sums up to 10.

I know adding math facts for 6 and 7 (highest sum 16).

I know adding math facts for 8 and 9 (highest sum 18).

Parent or Guardian signature

Figure 4.2. Math contract #2.

where his or her responsibility lay. If a child mastered a skill early, he/she could move on to some of the more challenging objectives that had not been highlighted. Information and suggestions were sent to each parent on how to work with the child to accomplish various aspects of the contract. In mathematics I suggested a number of manipulative activities to build understanding. One of these was a structure for using dominoes (see Figure 4.3). (Included in the appendix is a suggestion sheet to parents on other ways to use dominoes for number practice.) A parent was to sign off each of the tasks his/her child had mastered before returning it by the due date.

When the contracts were returned, I needed to make students accountable by checking them on their new knowledge, but with the first contract, report card conferences were just around the corner. Pressed for time, I again called on the fifth-grade teacher for help. Her students worked with mine, in a one-on-one setting, to do individual mini-assessments of the children. With the

Domino Equations

By _____

A Picture of My Domino	My Equation

Figure 4.3.   Domino equations.

overhead projector, I was able to give clear and simple directions to the two classes on how to use a simplified version of the contract as a check-off sheet. A great buzz of excitement filled the room as each pair found a place on the floor or at a table to begin the oral quizzing. The fifth-grade students were ever so kind in their encouragement as my students showed what they could do. Afterwards we had cookies and watched a movie together.

The mini-assessment was included in the students' portfolio of work for parents to see at conference time. After explaining how we had done our assessments, I was able to say, "According to this mini-test, your child is doing well with skip counting by fives and tens. He can still use more practice skip counting by twos. Does this match your experience?" I wanted the parent to know there was accountability for what he/she had signed off.

As a culmination to the contract, the class went on a field trip to Leatherby's Family Creamery where they had a tour and then ate ice cream sundaes. This was a reward for contract completion, but it was not an exclusive event. The rule was: "Everyone goes, or no one goes." I covered myself by saying that each contract was worth 10 points and that we needed all 200 points to go. Children who finished early could earn extra points for the class by doing more. This would boost our class total to compensate for the children who didn't quite make it. I was never exact about the points. I led children and parents to believe I had special records somewhere with a very careful count. I didn't. Some children did extra work. Some didn't quite complete everything. We all went.

Not Always Smooth Going

When working with parents on something that requires their extra work and commitment, it is not always smooth going. About three weeks into the first contract, I received a note from one of my parents: "As much as I admire your teaching methods, I must be honest. I simply haven't the time and/or the patience to go over all of Cynthia's assigned homework that you're asking. I understand your aggressive teaching of your kids. I'd just like to point out, for single working moms, for married working moms, moms with or without more than one child, non-working (paycheck wise!!) mothers, it's TOO MUCH! Please lighten up, as all homework pretty much demands my own participation to complete. Please be gentle. Help me to avoid any more guilt than I live with daily as it is." The moment I finished reading the note, I went to the phone and called Cynthia's mom. We talked, and I was gentle. I asked her to get the contract out and tell me what she *could* support Cynthia in doing. We cut back on the requirements. I said that whatever she could do was all that was being asked. The next time we saw each other, we laughed and hugged. By respecting Cynthia's mom's feelings, I had maintained the connection to this important supplier. Interestingly, when Cynthia's contract came in, almost everything was completed as set down in the original contract.

As I have said, not all the contracts came in 100 percent complete. I worked carefully with each family, always asking that they do as much as possible, but being sensitive to their problems. Abby's mother had cancer. Her first contract was late by almost a month. Abby's grandmother and older sisters helped her with it. Shelly's mom and dad were in a custody battle. She lived with her dad, who worked somewhere else and came home on weekends. So really, Shelly lived with the baby-sitter. I called the evening baby-sitter and began creating the vision with her that I would normally create with the parent. The baby-sitter worked with Shelly on the contract, and although it was late and incomplete, at least I had a home connection, and Shelly felt she had contributed to meeting our goal.

Because Lai Fin's parents could speak no English, I asked his brother Chan, a fourth-grade student, to come to my classroom after school to learn about the contract. He was enlisted to practice math skills and read with Lai Fin every night. The contract was never turned in, but I saw that Lai Fin was reading by the number of books he took home and brought back each week. This was an invaluable home connection for me. On Lai Fin's next contract, I asked only that he read every night. This was something he was doing already, so I knew he would be successful with it. Through Chan, Lai Fin was receiving the benefit of getting practice in reading, and at the same time was contributing to the class' success.

Mary's mother was having little success getting her to do her homework. To Mary's contract I added that she needed to turn in her homework every week. Two weeks into the six-week period, Mary had not turned in any homework. Mary, her mother, and I had a conference. I looked Mary in the eye and said we would not go on the field trip if she did not turn in her homework each week. I would not tell the class who had kept us from going; I would just tell them we did not have enough points, but she would know. It was a risky thing to do, because, of course, I would have to keep my word. We parted with Mary assuring me she would keep up her end of the bargain. Everything was fine up until the last week. When I checked the homework that Friday and hers was missing, I began to prepare myself for an aborted field trip. But Monday morning, there was Mary with a big smile, her homework in hand. We did a little joy dance.

We had gone on our field trip by the time I sat with parents at conference time. We looked at the whole experience from beginning to end. What did and did not work? What should we do differently? The next contract was created on the spot with the parent and child helping to make the decisions. In the areas of reading and spelling, we looked at what the child could presently do, based against standards set up by the school district and the California Reading and Literature Results Project. In math we went over what the child had worked on for the last contract. Sometimes I did on-the-spot check-ups so the child could demonstrate mastery, or show that more practice was needed.

Creating a New Contract

The old adage, "A picture is worth a thousand words," came to mind as I shared information using computer generated Class Action graphs. We celebrated the growth. Then together we chose new numbers, with me encouraging the child to press toward the standard. It went something like this: Me: "You can now read 100 words from the '200 Most Frequently Used Words' list. How many more do you think you can learn to read by the January deadline?" Child: "Thirty!" If I thought it was too high, I would say to the parent, "That seems a little high. Do you want to support Billy in doing this, or should we lower it a little?" The parent would then talk it over with the child, and they would decide what it was to be. If I felt the number was too low, I would encourage the child to go higher, but the number was never changed without the parent and the child supporting the choice.

In math we decided which things from the contract page still needed to be practiced, and then we went over another page, which had new math information to be mastered. A small clock with movable hands was included, as a way to practice telling time.

After the conferences, I was again on the phone with parents on a regular basis. "How is Hattie doing with her contract? Are there questions? Do we need to change any of the numbers? Is she going to be able to meet the deadline? Thank you for what you are doing to support your child's learning!"

For our next field trip, we went bumper bowling. With this kind of bowling, long spongy tubes are placed along the sides of the lanes, making it impossible to roll a gutter ball. One has not been properly entertained until one has watched young children bumper bowl! Some balls were bounced; some were released backwards; others took their operator with them down the lane. It did not matter because the children loved every minute. We whooped, hollered, and cheered whenever anyone hit anything. The time was gone before any of us wanted to leave. For the next month, the children relived the fun while they worked on math skills and solved problems that had a bowling theme.

Feedback from the Parents

I now wondered if the children and parents were tired of the contracts. Should we do something different? I sent home some questions to the parents. "How do you feel about the tasks you have been doing with your child at home? Has the way I have approached things worked for you? Are there things you would change? Are there things you have absolutely loved and would not change? What other thoughts do you have about the teacher–parent connection this year? Feel free to be honest." Many parents responded. They were overwhelmingly in favor of continuing the contracts. Here are some of their comments: "I really like your idea to do the contracts. A lot of times parents want to work with their children to help them keep up or in Dan's case to help him to catch up to the second-grade level. I had problems thinking of things that would help or I was unaware of where problems existed." "The reason I liked the contracts is that it told us specifically what to work on with our children and it offered the children an incentive to work hard. Also the way the contract is written it can be adjusted for each individual." "One other thing I thought was a good idea was the fact that there were a couple of contracts with deadlines. That was my incentive to work with Richard every day so we could meet the deadlines." "I feel like the assignments given to Susan have been worthwhile. I would probably be less motivated to help if it were 'busy work.' I also feel that my helping her has a direct and obvious effect on her learning. This too is motivational for me." "The contracts were not only fun and motivating, but they were specific and attainable with a realistic but firm deadline. I like clear expectations and deadlines."

For our next contract celebration, a parent helped me plan a pizza party that would take place in the classroom. We made a huge paper pizza for the bulletin board and each piece had a student's name on it. When children brought in their completed contracts, they decorated their pizza slices. We ate pizza in the classroom, and celebrated what we had accomplished. The pizza bulletin board was left up for Open House as a way of showing that the contract was a strong pillar of our learning program that year.

At the end of the school year, the class put on a thank-you program for parents and all others who had helped them with their learning. The theme of the evening was, "It takes a community to teach a child to read." The children read stories into the microphone. At the end, they sang "Because You Loved Me," and passed out carnations. The fifth-grade teacher, who had sent students as helpers, was there to receive hers. Lai Fin's parents did not attend, but Lai Fin walked to the program with his brother, Chan, who received one of the carnations.

Reflecting on Using the Homework Contract

During the summer I wondered about what it was that had made the Homework Contract such a powerful tool. I realized it gave me a purpose for establishing communication and a working relationship with the families.

Communication became the pipeline to my supply, that ever-important part of the educational system. Having all of us aligned to clearly defined goals was another part of it. When we came together by aligning with a common aim, a bond was formed that was energizing and enlivening. An added benefit that cannot be measured was the experience students had of seeing parents and teacher working in tandem to do everything within their abilities to help them to learn. It was easier for the child to align his/her own behavior with the efforts of his/her caretakers, when the aim at home and at school remained constant: Maintain enthusiasm while increasing learning. This, as opposed to what I have sometimes seen, where parent and teacher are at odds, with the child not sure which side to take. Most often it would be the parent's side, which means the child would feel a certain disloyalty if he/she embraced the learning opportunities presented by the teacher.

The homework contract also had no demotivating factors, no competition, no failure. Everyone was a winner. These are Deming cornerstones. Looking deeper, I saw that my own commitment was key. Commitment is a magnet that attracts commitment. As the children and parents did those things they said they would, as they kept their word, they began to experience a growing belief in themselves and the vision we had created together. Supported in keeping their word on completing challenging tasks, the seed called self-esteem began to sprout. There is nothing so motivational as a growing self-esteem. The taste is sweeter than candy, and more addictive. Self-esteem is the springboard for joy in the workplace, so valued by Deming.

I stayed with my class as they moved to second grade. In the fall, at Back to School night I asked the parents if they wanted to do the contracts again. The answer was a resounding "Yes!" My initial thought was, "*You* might want to, but I'm not sure I do. They are *a lot* of work." If that is all the further my thinking goes, I am looking at the trees, not the forest. Once implemented, I have twenty parents doing my work for me. Isn't this a magical leverage of time? How often have I been unprepared to use parent volunteers and classroom aides only to end up doing hours of work that could have been done by them? I think about my garden at home. Gary and Tanner spent time putting in an automatic watering system. Then they rototilled in a couple of truckloads of horse manure along with many wheelbarrow loads of mulch. After peas were planted, we all busied ourselves with other things while the rich soil, automatic sprinklers, and sun encouraged the little plants out of the ground. What a wonder to come out to the garden in a couple of weeks to see healthy plants in the once bare dirt.

As with the vegetable garden, the work I put into the home garden where my students are growing eventually saves me time and effort. Setting up to use the Homework Contract is not working harder; it is working smarter.

Summary

Educating a child is not something a teacher does alone. When the community becomes the hive that works together for the benefit of the child, the cooperation and camaraderie create a buzz of excitement, joy, and optimal opportunity for growth. I have used the Homework Contract as one way to create a common purpose among teacher, students, and parents. Looking at our desired outcome, I created concrete activities that parents and children could understand and do at home. Each contract was tailored to the needs of the child, as based upon assessment. Parent, child, and teacher talked together about adjustments that should be made to have the contract challenge, but not overwhelm the student. As individuals completed their contracts by the deadline, and maybe did extra for their classmates who didn't quite get everything done, the emphasis was placed on the success of the hive as a whole. Because all had given their best effort, all were included in the celebration activities. A teacher may choose to find another tool. However, the underlying secret for sweetening the partnership, no matter what tool is chosen, is the constant nonjudgmental communication.

Notes

1. Lee Jenkins, *Improving Student Learning: Applying Deming's Quality Principles in Classrooms* (ASQ Quality Press: Milwaukee, WI, 1997), 3, 21.

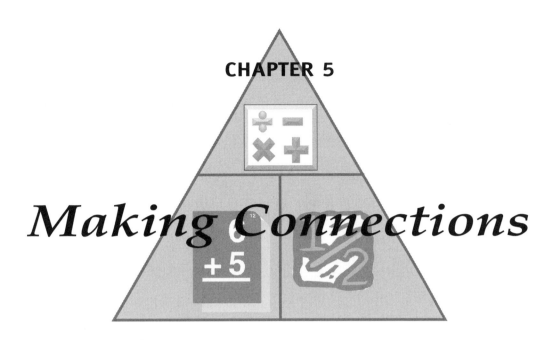

CHAPTER 5

Making Connections

In his book *The Web of Life,* Fritjof Capra reveals to the reader that life is systems inside of systems. He explains in scientific and mathematical terms that networks and patterns of organization can be seen in any aspect of life one wishes to explore—the atom, the Earth, the universe; the brain, body cells, social organizations. He goes on to say, "According to the systems view, the essential properties of an organism, or living system, are properties of the whole, which none of the parts have. They arise from the interactions and relationships among the parts. These properties are destroyed when the system is dissected, either physically or theoretically, into isolated elements. Although we can discern individual parts in any system, these parts are not isolated, and the nature of the whole is always different from the mere sum of its parts."[1]

Because connections are such an inherent part of everything, it is easy to overlook their simple wisdom; to overlook their value to the enhancement of the whole. I have come to realize that my job as a teacher is not to teach isolated bits of content, but to facilitate the children in discovering patterns and connections that already exist. It was a teacher that facilitated my first realization that we live in an ordered world. I will reflect upon this, along with some other realizations, and then use the concept of making connections to create a context for teaching.

Discovering Magical Order

I experienced a paradigm shift as I sat in my high school chemistry class looking at the periodic chart on the wall, realizing each element added one more electron. Up to that time, my view had been "We live mostly in a hodgepodge, random world. The things we learn in school have been invented so we can have a common language about the hodgepodge." The periodic chart opened up a new theory: "I do not need to invent order from the chaos. Order exists all around me. Learning is about discovering what is already there."

As time went on, other insights verified the magical order of things. One discovery was how much the solar system looked like a giant atom. How could it be? I was humbled (am still humbled) by this profound connection. I felt like an ant trying to see more than the hill I was on—which is also the way I felt as I read countless books describing the delicate balance called the web of life. What an intricate system of interconnections plants and animals create as they weave their lives together. Recently, I have come to realize that the DNA I am made of contains the same chemical parts as the DNA in all other living things. It's all the same ingredients.

When I became aware of the theory of quantum physics, again I had a shift in how I viewed the world. "The subatomic particles have no meaning as isolated entities but can be understood only as interconnections, or correlations, among various processes of observation and measurement. In other words, subatomic particles are not 'things' but interconnections among things, and these, in turn, are interconnections among other things, and so on. In quantum theory we never end up with any 'things'; we always deal with interconnections."[1] Not only are subatomic particles connected to each other, they are not separate from my observation of them.

A real find was the hologram. I read two or three accounts just to verify that such a thing really exists. "One of the things that makes holography possible is a phenomenon known as interference. Interference is the crisscrossing pattern that occurs when two or more waves, such as waves of water, ripple through each other. For example, if you drop a pebble into a pond, it will produce a series of concentric waves that expands outward. If you drop two pebbles into a pond, you will get two sets of waves that expand and pass through one another. The complex arrangement of crests and troughs that results from such collisions is known as an interference pattern." In short, "A hologram is produced when a single laser light is split into two separate beams. The first beam is bounced off the object to be photographed. . . . Then the second beam is allowed to collide with the reflected light of the first, and the resulting interference pattern is recorded on film."[2] To me, the process is about as clear as mud. I looked at diagrams, and I have no idea why it works, but the result is this: "Three-dimensionality is not the only remarkable aspect of holograms. If a piece of holographic film containing the image of an apple is cut in half and then illuminated by a laser, each half will still be found to contain the entire image of the apple! Even if the halves are divided again and then again, an entire apple can still be reconstructed from each small portion of the film. . . . Unlike normal photographs, every small fragment of a piece of holographic film contains all the information recorded in the whole."[2] A neurosurgeon named Pribram, while working at Yale in the 1960s read an article in *Scientific American* describing the first construction of a hologram. In working with the brain, he had been trying to explain how memory could be distributed throughout. The holographic model "hit him like a thunderbolt. . . . If it was possible for every portion of a piece of holographic film to contain all the information necessary to create a whole image, then it seemed equally possible for every part of the brain to contain all of the information necessary to recall a whole memory."[2] This information hit me like a thunderbolt, too. The implications go way beyond my present experience and understanding.

I've come to think that learning can be defined as what I do when I connect myself to the intricate web of relationships that already exist in matter. When I base my teaching on this truth, my role becomes that of helping students to make connections; helping them to get going on their lifelong jobs of discovering the magnificent order that already exists. By creating this context, students will be encouraged to begin searching for their own connections, creating their own awe-struck moments; climbing their own webs to new points of understanding.

Know, Knew, and New

To help me figure out how to help students make connections, I looked at learning itself, and noticed at least two ways I gain information and knowledge. One is in connecting something new to something I already know. The other is connecting something I know to something else I know, but didn't remember I knew. The step-by-step process used to learn mathematics is connecting what I know to the new. The value in an activity like The Hungry Bug, (in the appendix) is in the connection students make from what they know, 10 + 5, to something they do not know, 9 + 5. Another example of having children connect something new to what they know is by using the doubles. In an earlier chapter, I mentioned I taught students to sing "The Doubles." When children know that 7 + 7 is 14, it follows that 7 + 6 must be 13. However, it only follows if I assist the students in making the connection. When I first had students write about this connection, I could

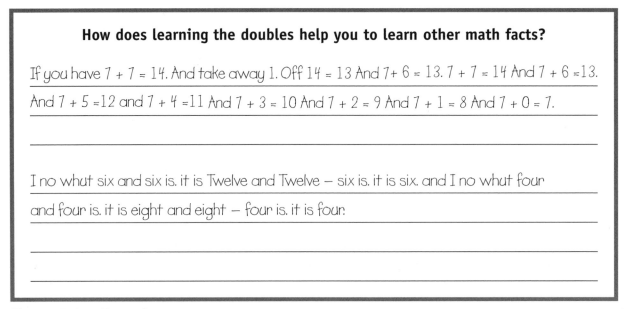

Figure 5.1.   Example answers.

almost see dots of sweat popping out on foreheads as they agonized. My experience is that a connection can come in a brilliant flash, but often it is preceded by some mental grinding (see Figure 5.1).

This kind of learning works in tandem with the other kind of learning—connecting what I know to what I already know. I had a recent hit-me-in-the-face experience with this. I have mentioned that I said "L-I-S," and the students said "T-E-N," when I wanted to interrupt and give a quick instruction. Sometimes I took a shortcut and just said the word, "Listen." We had done this routine more times than I care to count. Be assured the children had the spelling connected with the word. One day I spontaneously added the word listen to the end of the students' spelling test. About three fourths of the class missed it! When I talked to the children about it, the ones who missed it were incredulous. How could they have missed such an obvious connection? When I slipped the word onto a couple of later tests, all of the students spelled it correctly. Just by talking about it, students made what I call an in-house connection. Talking and writing about their thinking is the vehicle that allows students to begin to make connections—know to know, and new to know.

In one of the most eye-opening math assessments I ever gave, I removed many of the normal structures that are usually provided for children to show their understanding. I discovered that, for many, understanding was limited to the structure. In one question I removed all the calendars in the room and asked the students to draw the month of July, marking any special days they knew about. I got the most hilarious drawings of calendars I have ever seen. It was a good thing the drawings made me laugh, because otherwise I might have cried. For weeks and months we had done what all good elementary teachers do—our daily calendar work. Children could start on any day of the week at the top of the calendar, and trace down to show two weeks later. They could tell me there were about four weeks and 30 days in a month. They knew there were seven days in a week, and what those days were named. Or at least I thought they knew. The drawings showed the month of July with anywhere from five to 12 days in each week; anywhere from 23 to 60 days in the month (see Figures 5.2 and 5.3). Only four children drew a calendar I would call in the ball park. It became clear to me most of the students were peering through the calendar lines as if they were looking out a window, not seeing the panes because their focus was on something beyond. After their drawing experience, I replaced the calendar. Many children looked with new eyes, and saw the structure of the panes of the calendar for the first time!

I've never really recovered from the shock of this. How could the children see the seven days of the week across the calendar practically every school day since they were in kindergarten and not be able to show it in their own drawing? Clearly, I cannot find out what the children don't know, if I don't ask the right questions. I must go behind the scenes to search for ways to help them make connections.

When I created the assessment that included this calendar question, I thought I was just going to have a little fun. I did not realize the understanding that would come about for my own teaching. Now I see I was looking through my own invisible window panes. Here are some more of those panes (or should I call them pains?). With the clock in the room covered, children were asked to draw their own, and put the time they got up that morning. Again, hilarious drawings. Other questions were: (1) Write a three digit number. Then draw a picture of the number. (I thought children would do drawings of the base ten blocks we had been using all year. Some did. For the rest—more hilarious drawings.) (2) Draw a picture to show a multiplication idea. Write an equation to go with it. (3) Draw something you would buy, and put a price tag on it. Show the coins or bills with which to buy it. (4) Draw a picture showing what a fraction is. Write a fraction to go with it. (5) Write your own equations: two digit plus two digit—no carrying; two digit plus two digit with carrying; two digit minus two digit—no borrowing; two digit minus two digit with borrowing. (6) Make up the hardest math problem you know how to do. Then do it.

It might have been one of the most disturbing assessments I ever reviewed, but at the same time, it was by far the most entertaining. And I inadvertently found another way to help students make connections—know to know.

Ron Johnson invites students to get extra credit by showing in example and writing that they see their own errors in thinking. Brittany's Correction Sheet helps her think through what she needs to do differently in the future (see Figure 5.4).

Saterday	Monday	tosday	wisday	thresday	Friday	Saderday	Sunday
1	2	3	4	5	6	8	9
9	10	11	12	13	14	15	16
17	18	19	20	21	22	23	24
25	26	27	28	29	30	31	32
33	34	35	36	37	38	39	40
41	42	43	44	45	46	47	48

Figure 5.2. Calendar #1.

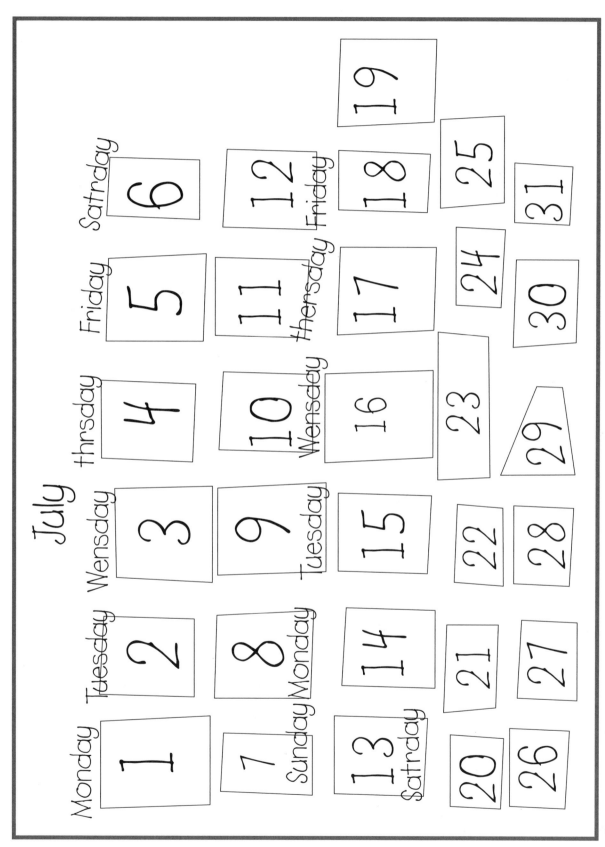

Figure 5.3.   Calendar #2. Artwork by Meghan Goulding. Used with permission.

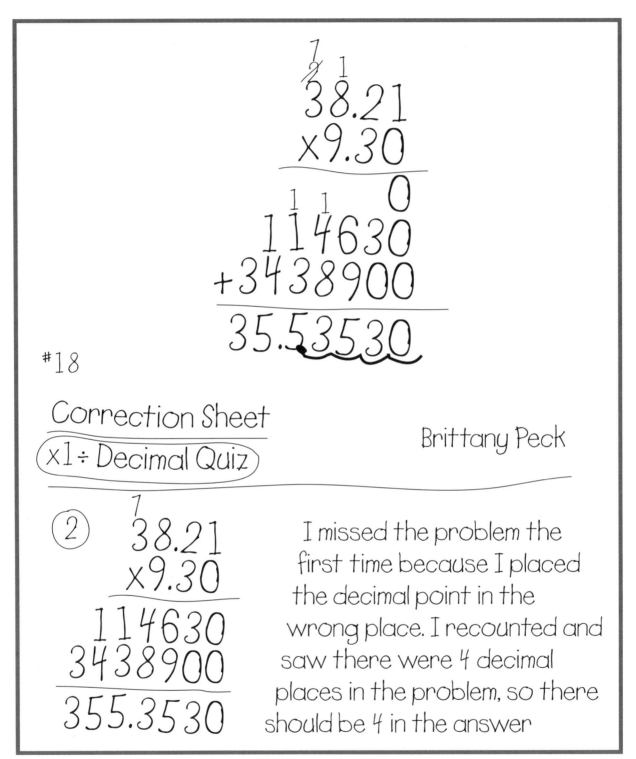

$$\overset{\overset{7}{\cancel{2}} \quad 1}{38.21}$$
$$\times 9.30$$
$$\overline{0}$$
$$\overset{1 \quad 1}{114630}$$
$$+3438900$$
$$\overline{35.\underset{\sim}{5}3530}$$

#18

Correction Sheet
~~Correction Sheet~~
(x1 ÷ Decimal Quiz)

Brittany Peck

② $\overset{7}{38.21}$
$\times 9.30$
$\overline{114630}$
3438900
$\overline{355.3530}$

I missed the problem the first time because I placed the decimal point in the wrong place. I recounted and saw there were 4 decimal places in the problem, so there should be 4 in the answer

Figure 5.4. Correction sheet. Task performance by Brittany Peck. Used with permission.

Connecting the Mathematical Strands

I see side benefits to using connections as an underlying structure when teaching. Looking at the eight strands of mathematics that the state standards say I should teach in a mere nine months, my first impulse is to laugh. Impossible! At least this is what I think if I'm not relying on connecting the strands. I am reminded of the way my mother fixed dinner when I was growing up. Our family rarely ate before eight or nine because my mother fixed just one thing at a time. When the carrot salad was finished, she put the potatoes on to boil. Taught separately, I would have to follow my students all the way through fourth grade to cover the material in each of the second grade math strands. This is why I must create a context of connection, connection, connection. A simple example of connecting two strands is pairing coins (see Figure 5.5). The children are asked to pair four different coins in as many ways as possible. With each pair they are asked to write the sum of the two coins or they could also be asked to find the difference. Matching the pairs of coins fits into the Discreet math strand. By finding the sums of the coins, students are also working with measurement and arithmetic, both in the Number strand.

Separating the study of mathematics into strands is a way to help students see and understand the parts that make up the whole. However, if the parts are studied in isolation, and no connection is ever made to the whole, it is like studying a bunch of bicycle parts, and never seeing that put together, they make a bike. I can learn about atoms and I can learn about the solar system, but only in their relationship to each other can I ever hope to understand the true nature of each. Since there are smaller wholes inside larger wholes, expanding out like Russian nesting dolls, children should work with the parts and wholes appropriate to their age and understanding. In this way, even at the kindergarten level, they can start to see that parts connect to wholes.

When I begin to think of ways to connect information between strands, as well as within the strands, I see the possibilities are endless. One day for their morning boardwork, I asked the children to use expanded notation as a better way of understanding two-digit and three-digit numbers. Some of the numbers I chose to work with were 365 and 52, these because we were learning how many days and weeks in a year. More time was not needed for me to connect these two things—place value and measurement. What was needed was me holding a thought: With every math lesson I teach, I look for the connections I can make. Whenever I dovetail different parts of the math curriculum, I feel smug—like I have fooled the curriculum gods who gave me too much to teach in the allotted time. If I can find a way to key students into the "trick" we have pulled, they also get excited.

Name _____

Pairing Coins

There are four coins: a quarter, a dime, a nickel, and a penny. How many different ways can you pair the coins? When showing the pairs, please give the sum of each set of coins.

Coin Pair	*Sum*

Figure 5.5. Pairing coins.

Connecting Myself to What I Teach

Students also get excited when they see I love what I am teaching. How I am connected with the material I am teaching influences how the children are connected to it. Experience is the strongest way I know to connect to something. This is true for my students as well as for me. I am connected strongly to what it feels like to sky dive, because I jumped out of a plane once. I remember clearly the fear I felt as I stepped onto the plane; the fear that grew as the plane took me up, farther and farther from the ground. I had to breathe Lamaze-style to calm my terror; but then I jumped, and I was free—falling, falling, falling. The "Zap!" of the opening parachute ended the wild one-minute rush, and I floated toward the patchwork quilt of fields until they grew from many to one. Do others want to hear about skydiving from someone who read about it, or from someone who experienced it?

A skydiver taught me about skydiving. It is not likely students will learn about pyramids from people who have gone to Egypt; but any time I can connect to information through my own experience, I become more alive when I teach it. This often happens when I attend a good teacher in-service. After attending an Orff music class in which we experienced the life of Harriet Tubman through drama and song, I put together one of the best teaching units I've ever created. Each year I relive my in-service experience with the children, who then join me in loving the memory of this courageous woman.

In theater, a good actress becomes the person she is portraying. As I watch, I am pulled into the drama. I forget my separation, and connect to the feelings the actress is experiencing. In the classroom I aspire to being this kind of actress, pulling students into living the learning. By participating in the Northern California Writing Project, I was able to begin seeing myself not just as a teacher of writing, but as a writer. I had a similar experience when I attended the CSU Chico Mathematics Project. Instead of just learning theories from a book, I was asked to think like a mathematician. When I become the writer, the mathematician, I burn with an enthusiasm that is contagious. I create a model students can emulate. Together we put on the drama called "Maintain enthusiasm while increasing learning."

Connecting to Other Curricular Subjects

The next connection I wish to make for the reader is the one that was the impetus for creating this chapter. Besides connecting to itself, math can be connected to other areas of the curriculum. Again, the curriculum gods are foiled as extra material is covered in a shorter time. However, it's not just extra material that gets covered. Each subject is changed as well as enhanced by its connection with the other. The whole that is created is greater than the sum of the parts. I will share some lessons integrating math with P.E., and following that, we will go to the grocery store for a math/health/social studies unit.

When I am doing data collection with students on what their favorite subject is, I often leave P.E. off so that I can see what their *next* favorite subject is. P.E. is a given. Why not use this as a vehicle for carrying math learning? The possibilities for integrating the two subjects are endless. One of the best lessons I have ever seen is the indoor P.E. stations created by the sixth-grade teacher Ron Johnson. Students work in cooperative groups to accomplish certain tasks while estimating, comparing, and calculating scores. I will give an overview and explain one of the activity cards as an example. The score sheets, teacher observation sheet, student self-assessment, roles of team members, and the other suggested activity cards are in the appendix.

Plenty of time is needed for these activities. Most of a rainy day could be used up. Students are placed in cooperative groups; each student in the group has a job. (1) The captain reads the station directions to the group. He/She makes sure that all members of the team complete each activity and the group rotates correctly through all the stations. (2) The noise monitor reminds the group to maintain a low voice. (3) The judge makes sure that all members are accurate and correctly recorded. (4) The checker makes sure that all group members leave their station neat and orderly, with all materials in place. One or two of the students can be the timers. The classroom is arranged so that students can move from one station to the next easily, and so they have room to accomplish the activity at hand. This might mean moving a jump rope station just outside the door in the hallway; or if the classroom is small and crowded, choosing activities that don't take a lot of space to accomplish.

Directions are placed at each of the stations. Students are given score sheets on which they must write the names of the activities to be performed. The teacher demonstrates each activity, and students use a pen to estimate what they can do at each station. The pen keeps them from changing their estimates after they get actual numbers. With the 5 Meter One-Footed Dash, the directions say, "Stand behind starting line. Balance on one foot. On signal to start, hop on one foot until you reach the finish line. Record number of seconds on your scorecard." With the pen, each student estimates the number of seconds he/she thinks it will take him/her to reach the finish line.

After practicing the rotation, each group is assigned a station and the fun begins. As each rotation is completed, students record their actual performance numbers beside their estimated numbers. When finished with all stations, team members compute their own personal scores, and then the combined team score. The score is the difference between the estimate and the actual measurement. A low score shows accuracy in estimating. Success in this activity is not dependent on athletic ability.

Before the stations begin, the students have been prepped with using stopwatches or personal watches for timing. Also, they know what is expected of their group, because they have looked at the Student Self-Assessment guidelines, questions which are answered when all the stations have been completed. Ron takes notes while the students are involved in their activities, and then at the end, he compliments each group in some personal way, emphasizing the same qualities on which the students assess themselves. Besides the information on the assessment sheet, students can personally acknowledge themselves for any of the parts they felt successful with—their athletic performance, their estimation, or their computation. As another way of recapping, Ron might ask, "What surprised you?"

Almost any elementary activity can be rewritten to fit the needs of a different age level. I did a scaled-down version of these stations for my second grade class. After doing this, I looked at what my students were able to do, and what they would need help with. I saw they needed to learn to use a stopwatch. And where sixth-grade students were able to compute differences easily, many of my students would need to use number lines, hundreds charts, and calculators. To deal with unfocused seven-year-old children waiting their turns, I modified the directions for the stations by having students *do* something with each difference they computed. For example, if the difference between a student's estimate and the actual amount was 15, he/she was asked to jump 15 times. To get the most direct feedback on the success of a lesson, I look to the children. As I watched them eagerly engaged in the math computations, in the physical activities, and in helping one another, I knew this math/P.E. activity was a keeper. (Sample activities are in the appendix.)

Running to Hollywood

The students were immersed in the P.E. stations for a day. How about being emersed in a P.E./social studies/math activity that goes on for days? One year when I taught a 3rd/4th grade combination class, we used the track to run to Hollywood, a destination chosen by the students. The social studies curriculum guide showed me I should be teaching students about our county and our state. One of our purposes was to become familiar with some of the terrain and city names, as we made our journey down the California map. Math, P.E., and social studies were integrated in this unit of study.

On the school track, 5 ⅓ laps was a mile. As each child completed a loop on the track, I handed him/her a popcicle stick as a record of that lap. So let's say that in a single day, students combined their numbers to come up with 200 laps. Two hundred divided by 5 ⅓ is 37 ½. The students ran a combined total of 37 ½ miles, headed south for Hollywood. Those who know how can figure this mileage with pencil and paper or on a calculator, but I was not ready to teach my 3–4 grade students the algorithm for division of fractions. How could I do it with popcicle sticks? One of the best ways to teach is for me to forget what I know, and go back to when I was a child. If I can remember or rethink the steps I went through to learn something, I can more easily guide students along the path. Many times I'm at the far end of the path, yelling directions to a child, who is barely within hearing distance. Because I can see the path closest to me, I am yelling, "Take the left fork in the path to cross the stream," but where he or she is stuck is at a mountain. I begin wondering why the child isn't getting to where I am. After all, I am giving such clear directions. What I'm not seeing is that my directions aren't fitting his or her predicament.

Using the popcicle sticks, we needed to figure out how to convert the 200 lap distance into miles. After talking about it, we decided to collect groups of 5 ⅓ sticks for each mile by exchanging single sticks for three parts of a stick. The sticks were not always broken evenly, but we all agreed to pretend they were. We then combined and recombined to get our total. I'm thinking that rather than breaking sticks, some sticks could be marked into thirds, in which case three groups of five whole sticks would need to be laid down with the marked stick to make 3 miles. Students' understanding of fractions, and later on, division of fractions, is enhanced by such real world activities.

Each day we used our total miles to determine the distance to move the stick pin south on the large California map at the back of the room. To help us determine mileage, we sometimes used the miles that were written in little numbers on the map, showing distance from one town to another; or sometimes we used the map key to estimate where to mark the next place our running took us.

Every time we had a spare minute, we were out running the track. There is nothing quite as enjoyable as working with a class that has a purpose. Sometimes I felt like I was driving a team of horses headed home for their feedbags. Hang on tight! Here we go! This was the year I wore a T-shirt that said, I must hurry up and catch up with the others, for I am their leader.

We wrote to the Hollywood Chamber of Commerce to get information about the city. They sent brochures and posters which we placed around the map. Every day when students entered the room, they gravitated to the California map to talk about places we had already been, compare miles run on different days, and count the remaining miles to Hollywood. I used this point of keen interest to include all kinds of math problems in our daily work. Many became adept at asking thinking questions for others to answer. Much of my math curriculum for that year revolved around this one activity. I could tell the students were alive with interest because the thinking sparks were almost visible.

Now that I am teaching second grade, I have thought about revamping this activity to try with younger children, but at the school I am presently teaching, there is no track. A solution emerged as I talked with Ron Johnson. He uses a modified plan that requires no track. A physical activity like running or jumping rope is chosen. The time it takes to perform the activity is linked with miles. Every few minutes is worth a certain number of miles, or a mile, given the distance that is being traveled. The map scale is used to figure how far the class has gone each day. In his sixth-grade class, if they are studying ancient civilizations, students can jump across Egypt. He suggests that if the curriculum is the Westward movement, students can run or jump across the United States. If the students' own state is being explored, they can use that for their travel. Ron uses the overhead projector to make a large permanent wall map for plotting each day's travel. The beauty of this activity is the high student interest that is focused around a rich, integrative unit, which includes math, P.E., social studies, and geography. The feeling of camaraderie the students get as they work together toward a common goal is the glue that holds the learning unit together.

Other Math/P.E. Connections

I recently visited Ron's classroom and he shared yet another math/P.E. lesson with me. He calls it Fraction Hunt—a treasure hunt with a twist. Students work in groups to decode crypted letters. Each letter is part of a word. When the words are formed, they give directions on some physical activity to perform, and where to find the next clue, which is usually way over on the other side of the playground. Running is an important part of the game. The prize is chocolate. As an example, I have included one of Ron's crypted activities in the appendix. It would need to be changed to fit the physical layout of the classroom and playground where it is being used.

Another running activity I have connected with math is the school jogathon that is put on each year by the Parents' Club to raise money for the school. Students collect pledges from family and friends. Contributors can either pledge a flat donation, or a certain amount of money for every lap the child runs.

On the day of the jogathon, children pin a paper with an identifying number onto their backs. Each time they complete a lap, a parent volunteer places a tally mark on the paper. After the jogathon, the parents' club mails out a form to donors, showing how many laps the child ran, and the donor sends in his/her contribution.

For the whole month before the jogathon, the class practices by running the jogathon route, but instead of using paper, we use the trusty popcicle sticks. Inside the classroom I have a jogathon practice poster in place—a grid with student names at the side and places for the date across the top. Each day after running, the students come in and enter their number of laps onto the chart. All through the month the students are counting and recounting their sticks. They are adding and re-adding their numbers on the chart. I do a lot of verbal questioning: "Today Nick ran 9 laps. Sarah ran 7. How many more laps did Nick run than Sarah?" "Uzi, Brad, and Lai Fin each ran 7 laps. How many laps did they run altogether?" After awhile, students asked the

Jump Rope

Name: _____

Estimate the number of time you think you can jump the rope without missing.

_____

Now jump the rope and count your jumps until you miss. How many times?

_____

What is the difference? (How close was your guess?) To find the difference you must subtract.

Show your work here:

Figure 5.6. Jump rope.

questions. When the jogathon is over, students glue their popcicle sticks together to make art creations.

Jump rope is one of my favorite P.E. activities. I am able to provide a good quality jump rope to each child for just $2. I start out by making a chart that says: I can jump rope continuously five times. Three or four other charts list different amounts—10, 25, 50. I have students count for each other, and when they have accomplished their goal, they sign their names on the chart. Periodically, they work with a partner to estimate how many times they can jump without missing; then they do the actual jumping. They figure the difference between their estimate and the number of times they jumped (see Figure 5.6).

Games like this keep boredom from setting in. Students are motivated to keep jumping. Jump, jump, jump. Count, count, count. Beat my last record; figure the difference. Try again.

I am convinced brain activity is connected to physical activity. Stephen Hawking, the wheelchair genius with Lou Gehrig's disease, might not agree, but I'm counting him as an exception. I have received many creative brainstorms and/or solutions to problems while out walking or riding my bike. As a matter of fact, physical activity has been used as a technique for solving problems connected with writing this book. I've walked to Hollywood already, probably more like to Egypt. My next learning experiment with students should probably be figuring out some way to compare their ability to perform problem solving tasks after physical activity, as opposed to solving them after a sedative period. I already know which way would bring the greatest joy to their hearts.

Bowling

Bumper bowling is a once-a-year physical activity I use as a reward for completion of the Homework Contract. Older classes would want to bowl without the bumper tubes, but looking at my students' final scores made it easy to determine that the bumper tubes were age-appropriate for second grade because, even with the tubes, students rarely bowled over 100 points. The bowling attendant put out the lightest balls the bowling alley had, but even so, the students' biggest challenge was dealing with the weight of the balls. I must use the school's video camera the next time we go. As I mentioned earlier, it was some of the funniest stuff I've ever seen kids do.

In preparation for the field trip, I previsited the bowling alley. Even though scores are now figured electronically, the attendant was able to give me a hard copy score sheet and rules for scoring. Coordinating our calendars, I set up our bowling field trip date. The first time I did this, the bowling alley generously gave me 10 old bowling pins to use in my classroom. Besides for P.E. activities, I have used them as a highly attended booth at our school's spring carnival.

Back at school, we began by bowling outside. We set up a limit line, and used a soccer ball. So students wouldn't have to stand in line for too long, half of the class jumped rope while half bowled until we switched. The children learned that all pins down on the first roll is a strike; all down with two rolls is a spare. For those students still struggling with simple subtraction, I continued to say, "Two pins are left. How many were knocked down?" We stayed simple—no accumulative score.

I created a paper in which students laid pennies on ten bowling pin circles and created their own subtraction problems, pulling the pennies off as they imagined they knocked down the pins (see Figure 5.7). To play the game

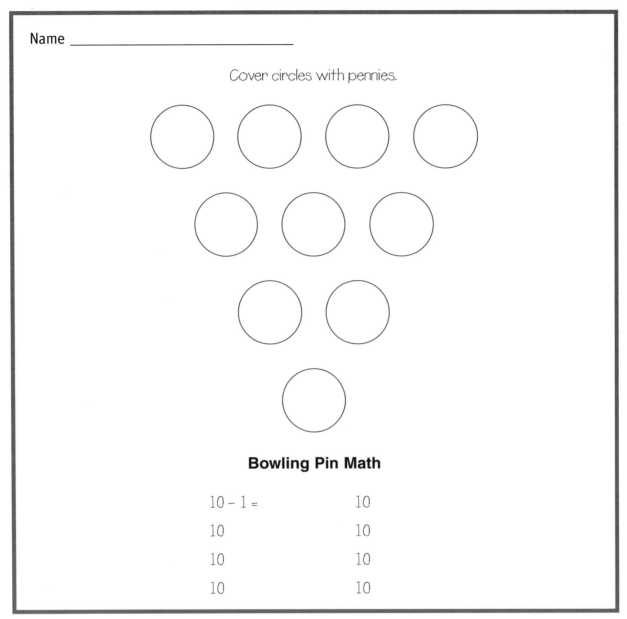

Figure 5.7. Bowling subtraction.

indoors, a score sheet was put on the overhead projector. (The outdoors activity was one appropriate for first grade. The indoor one was done with second grade children.) We found a place for a bowling lane, put masking tape down on the floor to mark the limit line and the placement of the pins. Two students were the pin setters. Other students took turns rolling the ball for two imaginary students I listed on the score sheet. As each student bowled, we did mental math to fill in the numbers on the score sheet projected on the white board. Older grade students would be able to each have their own score sheet to fill in as the game moved along.

After the field trip, students connected their bowling experience to other aspects of the curriculum by working on their Field Trip Notebook. The first page of the notebook had questions about the time we left and returned. Ours were simple: Draw hands on these clocks to show departure and return times. Write the times underneath. More demanding questions for older students could be: It took us 18 minutes to reach our destination. What time was it then? We returned 2 hours and 40 minutes after we left. What time did we return?

The next couple of pages were questions I created using student names: Meghan bowled a score of 57. Brandon bowled 85. How many more points did Brandon bowl than Meghan? Students were then asked to write one or two questions of their own, which they needed to show solutions for. I later typed some of these to use as quick activities at the beginning of the school day, giving credit to the students who created them.

Following the bowling word problems was lined paper. Students wrote about the field trip and drew a picture. At the back of the book was a 12×18 piece of paper that had been folded, with one side stapled into the booklet. When this paper was opened up, students had a place to draw a map of how they traveled from the school to the bowling alley. A map legend was required. The children were told ahead of time they would be drawing a map, so as we went along in the bus, I asked them to notice places we passed and streets we turned on. Naturally, the maps turned out to be more amusing than accurate. But that's okay. Being provided with constant amusement is one of the perks of teaching. And besides, when I am amused, my heart is lightened. Children respond willingly to a light heart.

The Grocery Store

Bowling is a one-time activity that, for me, is a keeper. Another is a field trip to the grocery store. The grocery store is a math teacher's dream laboratory. We are looking at a multi-age classroom with innumerable opportunities to use the math tools that have been practiced at school. Endless possibilities abound for estimating, comparing, weighing, and calculating. It's like having an exploratorium down the block, but for this one, there is no admission fee. However, the children do bring money: They shop for their parents. Because so many aspects of mathematics can be touched through this ready-made classroom, I begin a comprehensive unit of study in March and don't go on the field trip to the grocery store until May. In second grade, we would not benefit as much, going to the grocery store in September or October, because there are many beneficial skills to be utilized that we have not yet developed. Teachers of older grades have more flexibility in this area.

Even earlier than March we start working with decimals, and dollar and cent signs. Learning to add decimals on a calculator is a challenge for second-grade children, but in small groups and with persistence, the students can build proficiency. This is, of course, excellent practice for grades 3–6. Calculator activity books are available from math materials catalogs that have this kind of practice. Another activity we practice early on is estimating by rounding numbers to the nearest ten. This helps students keep from overspending when they get to the grocery store and have only a finite amount of money in hand.

To begin preparing the students, I start with the produce department and ask myself what kinds of things the students need to know to take full advantage of what is there. I purchased a scale which is able to hold small amounts of produce in its box-like container on top. A scale like this can be found in math catalogs for about $20. Seeing the actual produce sitting on the scale heightens interest, and gives the hands-on experience needed for understanding. Students begin to estimate and compare weights of similar and different sized foods. To guide their thinking, I ask questions: If one potato weighs ½ pound, and I add another potato of about the same size, what do two potatoes weigh? Three potatoes? If two bananas weigh a pound, about how much will six bananas weigh? How much more does a cantaloupe weigh than two potatoes? Once in awhile I throw in a curve to be sure the students are thinking: How much does a five pound bag of potatoes weigh? Something so simple can be quite perplexing to a child who is ready to do some really *hard* thinking. Teachers of grades 3–6 can ask these questions, and then more sophisticated questions. At first, I ask the questions. Eventually, students learn to ask the questions. When children can form intelligent questions, they have graduated from the information part of mathematics and are way into the knowledge of mathematics.

Using Food Ads

Because children are hooked in when they are seeing, touching, tasting, and smelling, using actual food is a good way to begin. Colorful food ads with all kinds of numbers plastered on them are the next best thing. Using the ads to make posters, I continue my questioning: Apples are 89 cents a pound. How much will you pay for two pounds? Cheerios are $1.53. Shredded Wheat is $1.20. How much more are Cheerios than Shredded Wheat? Students work these each morning while I am taking roll. As with the food weighing, after plenty of modeling, I start having students ask their own questions of my posters. They are guided from the beginning to state the fact or facts first, before presenting the question. Fact: Two bottles of catsup cost $1.58. Question: How much does one bottle cost?

Then comes the day they make their own posters. On the front of the poster is the statement and question (see Figures 5.8 and 5.9). On the back is the equation that answers the question. Or I have students put their solution

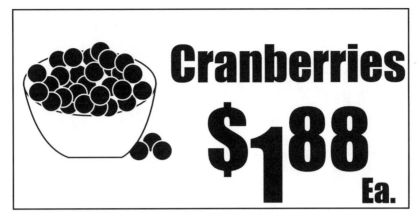

1 bag cost $1.88

1 dag of chip cost $1.68
Wats the difference of
1 bag of cranberries and A bag of chips

Figure 5.8.   Food ad problem #1. Story problem by Russell Adams. Used with permission.

if a bottle of Syrup is $1²⁹ and a box of Hot Cocoa is 89¢
How much would it be if you bought two bottles of syrup a one box of Hot Cocoa?

Figure 5.9.   Food ad problem #2. Story problem by Cassie Campbell. Used with permission.

on a strip of paper and enter it into an envelope they glue to the back. After I have helped the children reword their questions so they make sense for others to read, they enjoy going around collecting other student's solutions to their problem. When four or five strips of paper have been placed in the envelope, they dig them out to see if their friends were able to answer correctly. When checking them myself, I've found I don't need to do any figuring, because wayward equations stand out quickly.

Before giving the posters to me, students check their own posters against the clear directions that have been posted: (1) Write a fact or facts. (2) Ask a question. (3) Write an equation to answer the question. I am always amazed at how things that seem so clear to me can be so confusing for the children. Clear, visible directions can save a lot confusion.

A Classroom Store

With the weighing and the posters going on, what better time to have a classroom store? Students can bring in any kind of empty food container except glass. I always make sure milk containers have been washed out. After my classroom started looking like an overflowing recycling center, I learned to cut off the container contributions when a reasonable amount had arrived. Students have a grand time sorting the foods on classroom shelves. One group of children can sort and organize the food; another

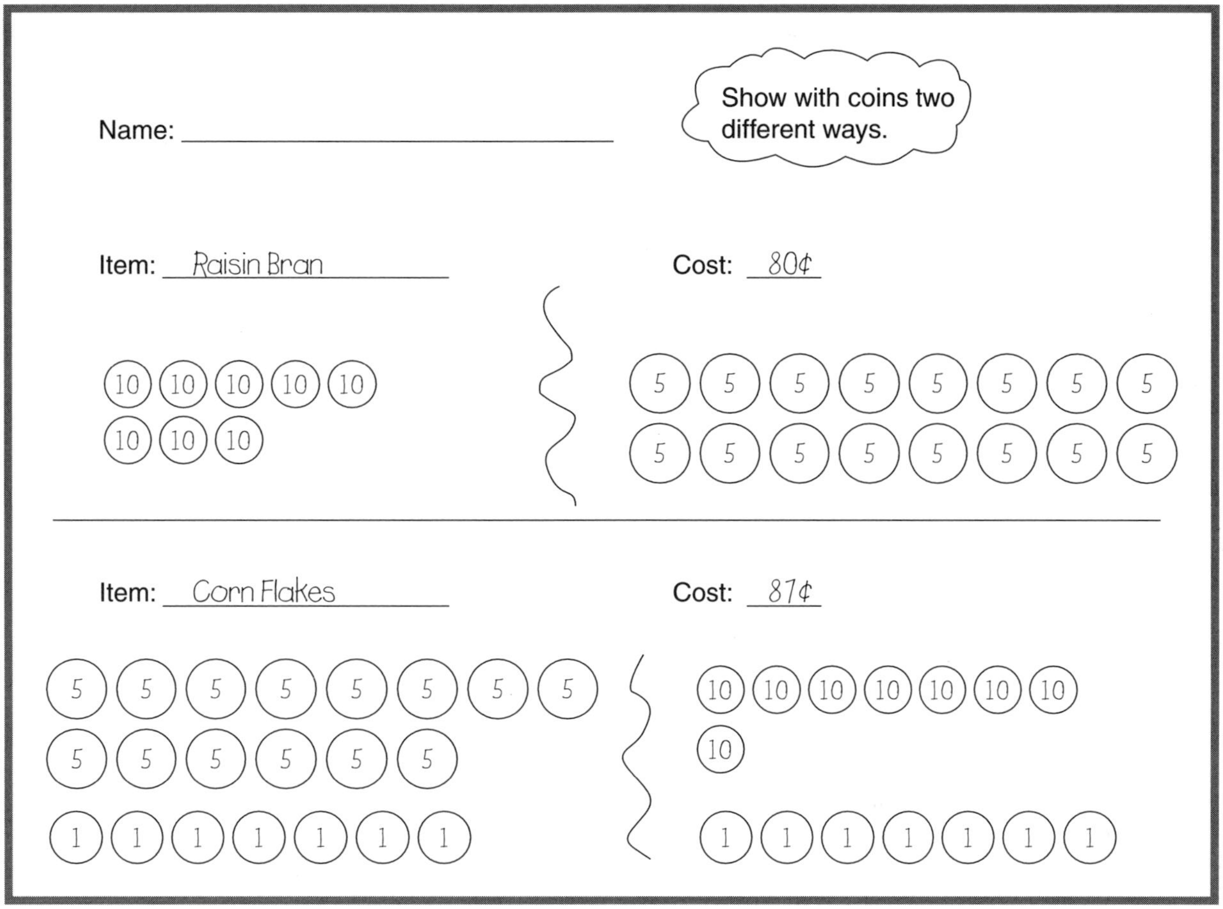

Figure 5.10. Classroom store worksheet example.

group can price. Still another can make big signs to lure customers into the store. Those who are pricing use a chart I make that gives ranges of prices to put on various categories of foods. For example, dairy foods: $.25–.50. We price much lower than actual amounts so that students may buy three or four items at a time and still be able to do the math required. Higher grades can use higher prices.

I usually have students go to the classroom store in small groups with their play money. Two or three checkers with calculators total each student's purchases. The checkers love to punch play cash registers and answer discarded phones if any have been donated. In second grade, students pay the exact amount so that no change is required. Older children can be asked to make change. Sometimes I use the store as one of three or four rotating math stations, with a parent guiding the children through their purchases. On rainy days, the store is a favorite hangout. A teacher colleague, Becky Luff, shared a worksheet that can be used in a math station (see Figure 5.10).

Learning to Count Money

As might be expected, a lot of preparation has gone into having the children be able to count up the money required to pay for grocery items. At the first of the year, some second-grade students cannot identify the names of the coins, and many do not know the values. Having a toy store where students use play money to buy personal items for themselves throughout the year is one way for them learn without even knowing they are. The teacher decides what kinds of things he/she wants to reward and then pays the students for these things with the play money. Examples are getting homework in on time, finishing class work, or coming into the classroom quietly. Any areas in which a teacher wants to change behavior are good ones to reward. Other teachers use the Math Their Way calendar activities to teach coins and their values. *Mathland,* a math text available in our district, teaches this through Daily Tune-ups. Overhead projector coins are used to work with the whole class. Dimes and pennies are a great way to teach place value while learning coin values. After students have learned to play Spin a Flat, a regrouping game in which students are racing to 100 using base ten blocks, we switch to dimes and pennies. It's the same game, but it feels different.

When I first start to teach children to count the value of a group of coins, I use a hundred chart. Children have learned to add on a number line. This knowledge is expanded to the hundred chart where they must now learn to track from left to right as they do in reading. Let's say the student has a quarter, a dime, two nickels, and three pennies. I suggest that starting with the quarter is an easy way to begin. We lay the quarter right on the number 25 on the hundred chart. We count 10 more for the dime. With 10 spaces across and 10 spaces down for the hundred numbers, most children have learned by now they can move from 25 to 35 by dropping directly to the number underneath. If not, they can count it out. We place the dime on 35. From there we count by fives, adding a nickel to the 40 spot and to the 45 spot. We lay down our three pennies by ones and we have our total of 48 cents! I can quickly check a student's thinking by looking where the coins are placed.

The money board is another good visual organizer. It can be used for learning to exchange one amount of coins for another. The board is made from a file folder that has been cut in half horizontally, which leaves the top and the bottom in two parts. This gives a 4 × 12 strip of tagboard to divide with five vertical lines. These vertical columns are needed for placing a dollar and/or any of four coins—a quarter, a dime, a nickel, and a penny. (The money board can be divided into six parts to allow for a fifty cent piece.) Headings are made at the tops with coin stamps—or play coins can be glued on. One cent starts at the right, and $1.00 is written at the far left. Here are a couple of ideas for using the money boards. Each student uses a pair of dice, or one student can roll for the whole group. A seven is rolled. The student gets a nickel and two pennies, and puts them in the proper places on the money board. Why not seven pennies? The rule is that a student must use the least amount of coins possible. Another seven is rolled. The student must exchange coins so he/she now has a dime and four pennies, put in their proper places on the board. I believe using small groups is the best way to successfully do this activity, especially at second grade.

This next activity is one of my favorites. It cannot be done until students are familiar with the names of the coins and their values. The lesson can easily be adjusted for grades 3–6. The children do an activity based upon the pattern set up in Judy Viorst's book *Alexander, Who Used to Be Rich Last Sunday*.[3] Alexander starts out with $1.00, and through one disaster or another ends up losing all of his money a little bit at a time. After I read them the book, children set up their money boards, starting with $1.00. As I read the book again, students break down and exchange coins as, along with Alexander, they lose or spend all of their money. For many second-grade students, this is quite difficult, so this past year, I had them backtrack on a hundreds chart—much easier! Exchanging the money is probably better for older students.

The children then write their own "_____ , Who Used to Be Rich Last _____" stories. I make up a little book with four pages, which has a place to write, and a place for pictures. Meghan wrote "Molly Who Was Rich Last Fall. Page 1. Molly had $100 last fall. Molly bought a car that cost $50. Good bye $50. Page 2. The price of gas went up and it costs $10 just to fill up her tank. Good bye $10. Page 3. Then Molly had to pay rent and that costs $30 dollars. Good bye $30. Page 4. Then Molly accidentally dropped $10 down the drain. Good bye all her money." Many of the children started with $1.00 like Alexander did in the original story. If I want students to work with coin values, I can ask that everyone do this. Instead of losing dollar amounts of money, they lose two dimes, then three nickels and so on.

After stories are written, students can be responsible for checking their own math, or they can exchange with someone else to do the checking. If a student starts out with $.85, and has four experiences of either losing or spending the money until it is gone, does all the lost and spent money add up to $.85? Sometimes I read the stories out loud, while students subtract each loss with a calculator. As with Judy Viorst's book, many of the children's books end up being quite humorous.

Another simple money activity I like is this: A line is drawn on a paper, about halfway across. With coin stamps the student stamps two or more coins to the left of the line and two or more coins to the right of the line. He/she writes the combined value of the coins for both sides. Then somewhere on the page he/she writes sum and difference. He/She adds the combined coin total of the left with the combined coin total of the right. Differences are also figured (see Figure 5.11). The amounts can be created to make

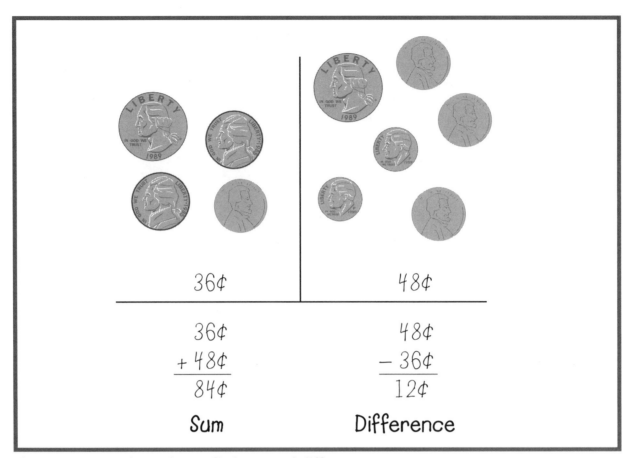

36¢

36¢
+ 48¢
84¢

Sum

48¢

48¢
− 36¢
12¢

Difference

Figure 5.11.   Using coins to find sum and difference.

the problem as simple or as difficult as the child wishes. A teacher can learn a lot about a child's understanding of subtraction by watching the child set up the problem. Doing subtraction problems from a workbook builds the skill—the information. Trying to subtract $.48 from $.36, getting frustrated, and then working through the frustration, can begin a whole new understanding for a child.

This is also a perfect time to integrate a healthy foods unit. In second grade, I do a unit on the basic food groups during this time. A fellow teacher, Cathy Reisfelt, developed a map of a grocery store that I like to use. Empty rectangular boxes have labels indicating frozen foods, produce, dairy foods, and canned goods. At the side of the map is a list of foods. Students sort the foods into their proper places by writing them into the map sections where they belong (see Figure 5.12). This creates the awareness that things in a grocery store are not just haphazardly and randomly placed, but have a planned order, a concept that is reinforced when students set up their own grocery store in the classroom, and later visit the real grocery store.

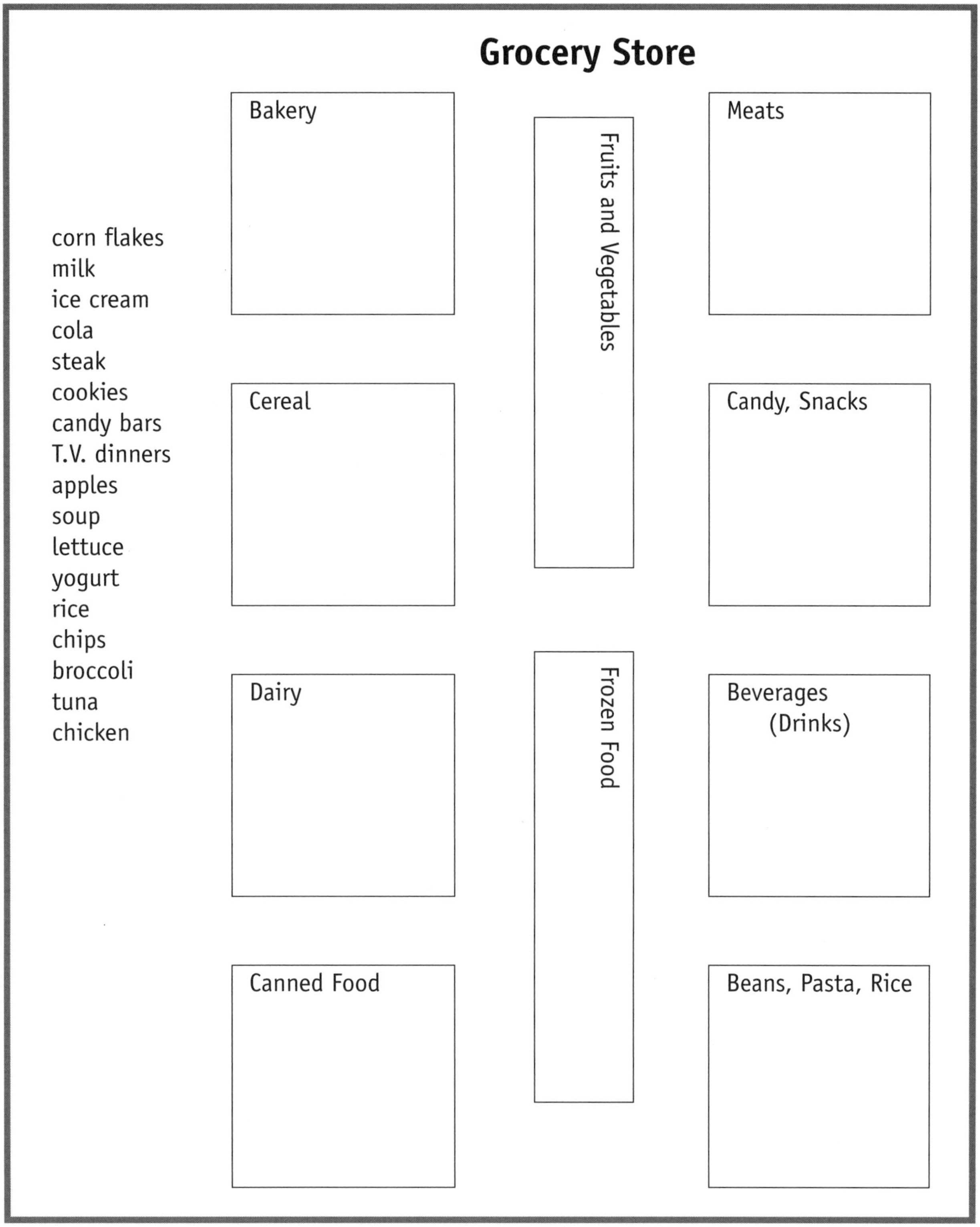

Grocery Store

Bakery

Meats

Fruits and Vegetables

corn flakes
milk
ice cream
cola
steak
cookies
candy bars
T.V. dinners
apples
soup
lettuce
yogurt
rice
chips
broccoli
tuna
chicken

Cereal

Candy, Snacks

Dairy

Frozen Food

Beverages
(Drinks)

Canned Food

Beans, Pasta, Rice

Figure 5.12. Grocery store map.

Getting Ready to Go to the Grocery Store

As the time nears for the actual visit to the grocery store, I call the manager to tell him/her the day and the time I am planning to show up with my class, so he/she can be forewarned. A note has gone home to the parents: "We will be going on a field trip to the grocery store from 9–10 A.M. Along with the permission slip, children need to bring $3–5 and a grocery list. Please send the money and grocery list in a sealed envelope with your child's name on it. Your grocery list *should have more items listed than the child has money to buy.* This way children are required to estimate and do mental math to choose from the list, items that add up to something less than the amount of money they have. Suggested items: toothpicks, cans of vegetables, soup, or fruit; fresh produce like apples, potatoes, oranges, bananas; Jell-O, soap, mustard, catsup, paper plates, napkins. Any food your child buys *cannot be eaten by him/her that day.* All groceries are to be taken home along with any change received. *Children will be allowed to buy only those things on the list.* When we get back to the school, we will check our sales slips with the calculators and count our change. *These are items that you should not put on your grocery list: anything in a glass jar, frozen foods, or foods that require refrigeration.* Please let me know if you can help with this field trip. A ratio of 2–4 students to each adult allows for maximum learning."

On the field trip day, I anticipate that some children may not bring a grocery list and money. I take care of this problem by having some grocery lists of my own ready. I give the children these lists so they can shop for me. Before getting on the bus, we review proper grocery store etiquette. When we arrive at the store we collect outside and I group students with parents. Each adult is handed an instruction sheet:

At the Grocery Store

"Keep your group together. Children can help each other to find items. This way students learn not only about their own items, but others' as well. Help children to estimate, by rounding off–69 cents rounds to 70 cents–so they can mentally add up items they are purchasing. We hope they have more things on their list than they can buy. This way they must pick and choose. Remember, no refrigerated or glass items. Be aware that tax will need to be added to some items. The children do not know how to figure the tax. Children may not buy something that is not on their grocery list. Also, they are not allowed to eat anything they buy. All items and all change go in the check-out bags, along with the sales slip. Names should be put on the bags. When all grocery items have been collected, go to the produce section to have the children use the scales. They can estimate and then weigh different fruits and vegetables. If one pound costs 69 cents, how much will two pounds cost?

"Remind the children to be polite to the other customers in the store. Have a good time! Thank you for helping us with our learning!"

Back in the classroom, we spend time in the afternoon becoming familiar with the sales slip. We first check to see that we have all the items we were charged for. Using calculators, we add the amounts on the slip and add the tax. Do our totals match? What happens when we subtract the total from the amount we handed the cashier? Do we have the correct change in our bags? If there is time, I ask students to write about their experience. On another day, we write letters to the manager of the grocery store thanking him/her for being patient with the 20 little students who rushed the aisles of his/her store.

Connecting Math and Literature

There are volumes that could be written about connections to be made between curricular areas. Briefly, I would like to outline a math/literature connection my students enjoyed, and beyond that, leave the reader to continue the connection journey on his/her own. To begin this unit, I found four storybooks that contained different numbers of people and animals: *The Great Kapok Tree,* by Lynne Cherry[4]; *The Fat Cat,* by Jack Kent[5]; *Farewell to Shady Glade,* by Bill Peet[6]; and *The Big Red Barn,* by Margaret Wise Brown[7]. After I read the books to the children, they worked in pairs to make bar graphs of the number of animals they found in their books. For example, *Farewell to Shady Glade* has one raccoon, one skunk, two possums, six rabbits, and five frogs. I conferenced with each pair of students on how they could make the lines for their graph. Children were encouraged to use different methods for filling in the squares: stickers, stamp and stamp pad, small geometrical drawings, pieces of construction paper, stencils, or coloring. Choosing a part of the story to recreate, students made a big pastel drawing to accompany the graph. They were proud of their results, and well they might be. The effects were stunning.

An easy way to connect math with storybooks is to use any of the many books that have been written using some aspect of math as the theme. A collection of my own such books are listed in the references.

One obvious area I am not writing about is the connection to be made between science and math. For further reading, I recommend Jeff Burgard's[8] and Shelly Carson's[9] books. They have each written accounts that break new ground in connecting curricular areas, one in science, *Continuous Improvement in the Science Classroom*[8] and the other in social studies, *Continuous Improvement in the Social Studies Classroom*[9]. Writing is another curricular area that I have woven throughout the book, showing how the children use it as a tool for thinking mathematically. May the reader have fun with some of these connections, and go on to make many more of his/her own.

Summary

In this chapter, I marvel that the fabric of our lives is woven with magnificence, as I notice countless networks of patterns that nest inside more countless networks of patterns. I realize my role as a teacher is to facilitate my students' discovery of the order that already exists. I give examples that show children can be guided to connect what they already know to something new. And that they can come from a different angle at what they do not know they know, to bring it alive from its dormancy. Students are inspired when I, as the teacher, am strongly connected to the material I am presenting. In teaching mathematics, the whole is greater than the parts, when I connect the strands and when I connect mathematics to other subject areas—P.E., social studies, geography, literature, and writing. Not mentioned, but an important part of the fabric, are social connections. Children in the classroom and teachers in a school, when connected in purpose to others, have a feeling of belonging. Not only does this enhance the climate of cooperation and learning. I think the strongest human urge is the desire to feel connected—to each other, to the Earth, to ourselves, to what we do. Deming's vision of a system where all are valued and respected as they work toward a common aim is profound because it sits at the core of this urge.

Notes

1. Fritjof Capra, *The Web of Life* (Bantam Doubleday Dell Publishing Group, Inc.: New York, NY, 1996), 29–30.
2. Michael Talbot, *The Holographic Universe* (HarperCollins Publishers, Inc.: New York, NY, 1991), 14–17.
3. Judith Viorst, *Alexander, Who Used to Be Rich Last Sunday* (Simon & Shuster Children's Publishing Division: New York, NY, 1978).
4. Lynne Cherry, *The Great Kapok Tree* (Harcourt Brace and Company: Orlando FL, 1990).
5. Jack Kent, *The Fat Cat* (Scholastic Book Services, Inc.: New York, NY, 1971).
6. Bill Peet, *Farewell to Shady Glade* (Houghton Mifflin Company: New York, NY, 1966).
7. Margaret Wise Brown, *The Big Red Barn* (Harper and Row Publishers, Inc.: New York, NY, 1989).
8. Jeffrey J. Burgard, *Continuous Improvement in the Science Classroom* (ASQ Quality Press: Milwaukee, WI, 2000).
9. Shelly C. Carson, *Continuous Improvement in the History and Social Science Classroom* (ASQ Quality Press: Milwaukee, WI, 2000).

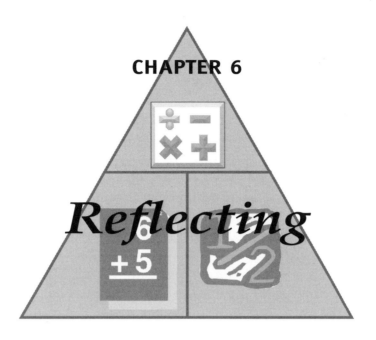

CHAPTER 6

Reflecting

To keep my friendship alive with a high school friend who moved, I wrote long, involved letters in which I shared just about every thought I ever had, and some that I didn't have until I started writing. The funny thing is, because of the sharing, I felt closer to Lindsey than before she moved. I never saw her again, and as time went along our letters dribbled to a stop, but I think of her now, as I reflect upon writing this book. I realize I have formed a friendship with the reader. Since I do not receive responses in the mail, the reader is what I would call an imaginary friend. From the perspective of the speaker or the writer, the magic of an imaginary friend or a pen pal comes from the perceived listening.

In my letters to Lindsey, I began by reporting newsy events, but the process of writing activates thought, so soon, I was responding to the ideas generated by what I wrote. Writing is like that. It lets me talk with myself about my own ideas. With the book, first I was writing about what I had done, but then as I wrote, ideas came to me about ways to improve upon what I had done. I began wondering if I was forming the thoughts, or if the thoughts were forming me. This phenomenon continued as I was thinking about the conclusion for the book. I named the chapter "Reflecting," before I knew all that it would contain. I intended to pull back, look at what I had written, look at my school year, and glean some wisdoms to share with my imaginary friends. My thought processes eventually took me to what the students had learned, and I started thinking about how silly it was to do any learning activity without afterwards reflecting upon what was learned, what should be kept, and what should be changed. I wished I could tell the reader that each day, after the children were gone, I had reflected upon the happenings, the learnings, the high points, and the low points. My next thought was, "Well, why not do it now? There's still some time left." I got busy creating a question sheet I could use to spur my thinking: What was the highlight of the day? The low point? When did the children show the most enthusiasm? When did they seem to drag their feet? When could I have listened better? Which children were least engaged in their learning? Why? What parts of lessons should I keep? What should I change? Did I help the children to make connections? Were the needs of the very capable as well as the struggling students met? What could I have done to enhance their participation and learning? Did I put the volunteer and aide help to best use? Did the physical arrangement of the room, and the ways I grouped the children enhance learning? To keep from robbing myself of a reflection time, I decided not to straighten the room, run things to the office, check papers, begin next day's lessons, or go to meetings until I have reflected upon the day.

I usually try to be as efficient as possible when school is out. Rush here. Rush there. But what good does it do for me to go somewhere fast, if it's not where I want to go? Reflecting for a few minutes after school each day is looking out the window of the train to see if I need to get out and lay new track. Instead of stealing my time, reflecting is giving me time.

I am reminded of the 10 minutes of reflection time my daughter, Shawni, and I took back in October about the family Christmas traditions. Each Christmas, for over the past 30 years, the extended family had gotten together to exchange gifts and have dinner. By mid-morning the front room where we converged looked like a cluttered department store warehouse. As the day went on, friends from nearby arrived and in the late afternoon, we sat down to our Christmas dinner together. As the make-up of the family changed, an unsaid mandate kept us following these same traditions. When Shawni and I talked we realized what we now preferred was a smaller gift-opening, each at our own homes, and then a family get-together for dinner. At Thanksgiving, we bravely announced we would not be buying everyone gifts for Christmas. Immediately the room filled with the buzz of people's responses. The end result was a decision to come together for dinner, with each person bringing one $20–$30 gift. We would play the crazy gift-opening game where a person may take an unopened gift, or one that has already been opened. Not only did the 10 minute reflecting save Shawni and me hours of time, and purse loads of money, everyone appreciated the change. The jolly laughing we did after dinner as we passed gifts from one person to another, was not only from the amusement brought about by the game; I think everyone was secretly laughing because they wouldn't be paying off credit cards until the middle of July.

The daily after-school reflecting has been immediately rewarding. When I reflect, I don't respond to each question I have written. To do that would mean an hour or two of reflecting. But the questions help me dig out what is not otherwise obvious. As reflecting becomes more and more of a habit, hopefully I will find myself spontaneously doing it throughout the day. To some extent, I do it now, but too often I find that I have gone through a whole day, maybe a whole week, without asking myself if what I am doing is taking me where I want to go. One of the things that suppresses the desire to reflect is the realization that such thinking will bring about change. Teaching—and living life in general—is a continual tug between status quo and change. Dead is the extreme condition of a constant status quo. Courage may be required, but reflecting can bring changes that freshen a stale classroom (or a stale Christmas). If I can allow myself to ask the forbidden questions, there is nothing that says I must change. I choose. However, once a forbidden question has been asked, the mind begins immediately to look for resolution, and change becomes a natural outcome.

Reflecting on what happened yesterday or today in the classroom is connected to a longer time tunnel stretching back over the past months and year. I asked myself: Which activities will I repeat next year? What will I change? As I looked into the crystal ball of the past I saw my After School Club. When students are struggling, and understanding comes in its own sweet time, I sometimes get discouraged. Did the time I spent with these children make a difference? Was it worth the effort? When I looked in the crystal ball, I saw the students feeling good about themselves. I saw parents working and having fun with their children. I saw students, teacher, and parents working cohesively, a team of people with a purpose—a worthy model to perpetrate. And I saw the children loving to do math. Yes, the After School Club was a worthwhile endeavor. As these children move on through school, they are more likely to stick with the challenges if math is something they love. They will continue to build their own math picture from puzzle pieces they began playing with in second grade.

After reflecting about the After School Club, I looked at the Math Enthusiasm wall chart. In the push to get results, it is easy to forget that self-esteem and joy are at the heart of the desire to learn. With all that we had done this year, had the students maintained their yearning for learning? I had shining results. I decided to see what the results looked like on the Class Action scatter matrix. Beautiful! Here's a Chinese Checkers game I won (see Figure 6.1).

This next year I will not be conversing with readers, telling of my foibles and successes, but I'll probably continue to write as a way of talking with myself about what works and what doesn't. I get to begin the school year standing on steps I have built. I discovered problem solving *can* be taught concurrently with math information, something I had wondered about at the beginning of the year; but instead of experiencing fear and discouragement

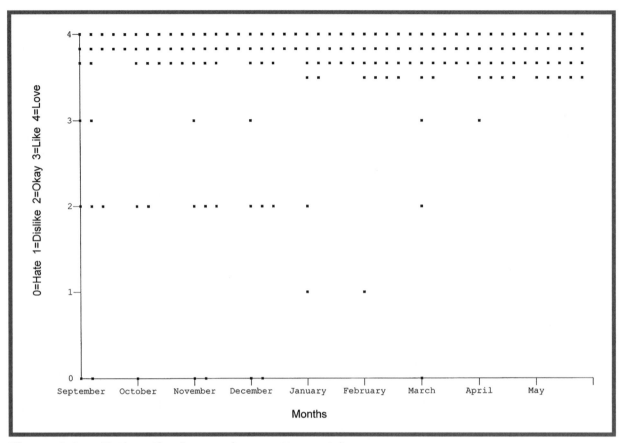

Figure 6.1.   Math enthusiasm—class scatter matrix.

as they try their first problem, students will have participated in a carnival of activities that leads them to the problem they will be doing by themselves. I have much to learn in guiding children through the problem-solving process. An abundance of books and other resources abound. However, without enacting a PDSA model, the good ideas I learn about will remain as written words on the page—hidden treasures.

The Pareto Chart

With the Enterprise Weekly, the step I have to stand on is the Pareto chart. By plugging in the number and kind of errors students made on the culminating 40-question test, I have a graphic view of the children's strengths and weaknesses. "The purpose of a Pareto chart is to separate the significant aspects of a problem from the trivial ones . . . What makes a Pareto chart different from other bar charts is that the data are arranged in descending order with respect to its frequency of occurrence. The incident or event that

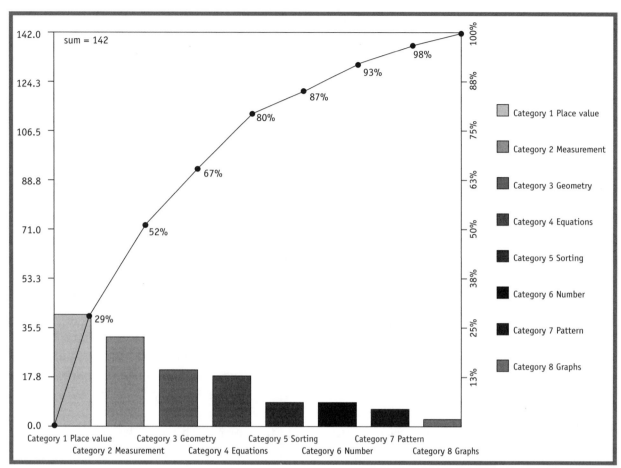

Figure 6.2. Enterprise Weekly final test.

happens most often is on the left, descending to that which occurs least frequently.["]1 Besides having an immediate visual look at the largest category of
concern, a percentage line gives me accumulative percentages as I move to
the right on the chart. By looking at my Pareto chart, (see Figure 6.2) I can
see that 52 percent of my problem areas are in place value and measurement.
I lumped linear measurement, time, and money into the measurement category for the Pareto chart. I can think of place value being even a bigger area
of concern when I realize students often have problems with money because
they don't understand place value. If I add in geometry, I am looking at 67
percent of the students' difficulties with math information. This gives me the
place to focus on for improvement. Next year, I can continue working with
equations, sorting, number, pattern, and graphs, but for continuous growth
I would be wise to put a good share of my energy into place value, measurement, and geometry.

Another teacher told me she felt her work with second-grade students in
other bases did more for the students' understanding of place value than

anything she had previously tried. On a limited basis in past years I've taught place value using other bases. But I thought it gouged out too much of the school year, so I quit. Looking at my Pareto chart, I see I must retry teaching in other bases. If I can flatten down those two tall bars on the left, who cares how much time was gouged? I can see my work will never be done. As soon as I finish one learning experiment the next one is in the hopper: Does work in other bases enhance second-grade students' understanding of place value? I better get the wheels of my PDSA cycle oiled. It has a new place to go.

The Fishbone

I included students in my year-end reflecting by asking the question, "What are the things we do to be better math students?" They each wrote ideas on small sticky notes and placed them on a large piece of butcher paper. The following day we read what we had, and then began a brainstorm of new ideas. On the third day, I had the children take turns sorting their sticky notes onto a large fishbone diagram which was made on another piece of butcher paper. After separating the mush of their mathematical year into the distinct aspects that contributed to their learning, students were ready to reflect upon the vehicles that brought them to their present mathematical understanding. The fishbone is called a cause and effect diagram because it helps answer the question "Why?" In this case, it helped students connect where they were now with those things that had brought them there.

As I took the contents of the butcher paper fishbone and transferred it to the Class Action computer program (see Figure 6.3) Gary said, "I don't see why you are making such a big deal about the fishbone. What makes it any different than the written diagram that has been done for years, where you have one, two, three, four, with a, b, c under each?" When I tried to respond, my tongue wouldn't work. Although I felt there were benefits to the fishbone diagram, I had not tried to explain them to anyone. So after Gary's question, I pondered; I did some reading; now I am ready to discuss the value of fishbone diagrams with my readers. I thank Gary for the question. When he reads the book, he can get his answer.

The explanation is quite simple: As I said earlier, in referring to student run charts, a picture is worth a thousand words. I can have someone tell me, or write down the directions on how to get to his or her house: "Take the freeway south to Riverside Drive; go east on Riverside Drive to North Street. Turn left on North Street and go across the river and up to the light. Turn right on Dersch Road and drive until you come to Prairie School, etc. etc." Or someone can draw a few lines on a page, indicate which direction is North, and I have an immediate visual picture that sticks in my mind. The same can be said of

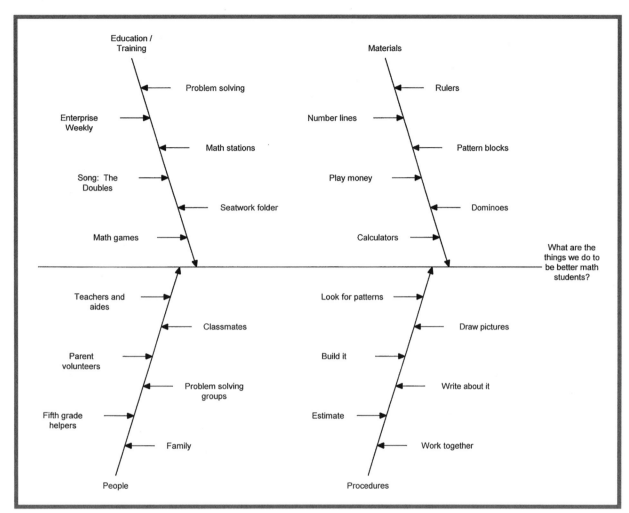

Figure 6.3. Fishbone diagram.

house plans. Imagine giving the house builders a written description of how to build the house. The architectural drawings save thousands of words, and immediately give more clarity. Another example is the monthly calendar. The grid we put our days and weeks on is a pure construct of the brain that makes it possible for me to know I will be able to attend a class reunion four months in the future without having to draw tally marks in the dirt.

Quality tools like the Pareto chart, run chart, scatter matrix, histogram, flow chart, and fish diagram are architectural drawings for the mind. Once seen, they are simple, and so their value may be overlooked. I wasn't too excited about the fishbone diagram until the day I drew one to solve a problem of my own. As I reflected and tried to make connections, the drawing nudged my mind into a higher gear, and I literally solved the problem right while I was filling in words on the bones. I want to do this for students—help them slip from the lower to the higher gears of the brain—the gears that

Enterprise Math Concepts

In this test, students solve a variety of math problems. These problems assess students' abilities to compute, measure, and show other math understanding. There are 40 items on this second-grade test. In order to meet Standard (grade level), students need to have 33–36 problems correct.

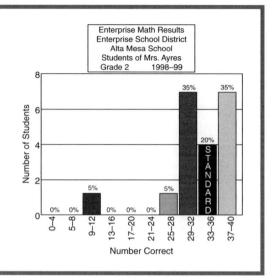

Figure 6.4. Enterprise math results for Mrs. Ayres' class. Used with permission of the Enterprise School District.

Bloom's Taxonomy[2] calls: knowledge, comprehension, application, analysis, synthesis, and evaluation. When I say I want to create continual improvement, what I really mean is I want to move students continually into the higher gears of brain activity. The fishbone and other such charts are brain tools—simple but profound drawings to help the brain move up the thinking ladder to find solutions.

Delving Further into Test Results

After looking at my own class' results of the end-of-the-year Enterprise Weekly test (see Figure 6.4). I looked at its results compared to our school's second grade, (see Figure 6.5) and to the Enterprise District's second grade (see Figure 6.6). I saw that I had 55 percent of my students at or above standard, and that 63 percent were at or above standard in our whole second grade. When the category just below standard is added in, the numbers are 90 percent and 84 percent, respectively. In looking at the district's second grade as compared to mine, I was excited until I realized my higher bars indicated the percentages of students who had missed the question. Wanting my scores to be better, my mind began to search for justifications. I realized any time I begin justifying in this way, I am separating myself from my purpose in looking at the data. The process starts being about me instead of about the children, and about my bigger purpose, which goes way beyond my own ego. Openly looking at the data to extract meaning brings useful insight and knowledge that guides me to that bigger purpose. What does

Enterprise Math Concepts

In this test, students solve a variety of math problems. These problems assess students' abilities to compute, measure, and show other math understandings. There are 40 items on this second grade test. In order to meet Standard (grade level), students need to have 33–36 problems correct.

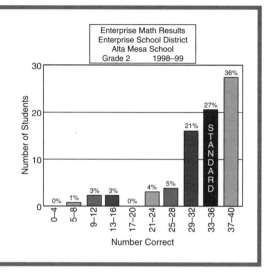

Figure 6.5. Enterprise math results for Alta Mesa School, grade 2. Used with permission of the Enterprise School District.

this data suggest, considering my goal of continuous improvement? (1) I have much to learn from my second-grade teaching partners. (2) I must decide if I should continue my one-year experiment of tracking student data using the PDSA model.

Before answering the second question, I looked at my class' Standard Achievement Test, 9th Edition (SAT9) scores. The Enterprise School District did not give a culminating second-grade problem solving test. When I asked, I was told we would use the Standard Achievement Test, 9th Edition (SAT9), given state-wide at the end of the year for that information. This would be possible because math on the SAT9 is broken out into procedures (information) and problem solving (knowledge). I am not fond of the SAT9, or any standardized test for that matter, because the percentile score, which is the reporting measure that gets the most attention, requires winners and losers. Interestingly, the state of California is now doing a big push to get all schools above the 50th percentile. This, of course, is ludicrous, since by the nature of a percentile score, 50 percent of the student population must be below. I realize this is an effort to have California stand proud among other states in the nation, but someone, somewhere has to lose. My own feelings aside, the SAT9 is a required test. Scores are sent home to parents and published in newspapers. Their numbers are something for which I must be accountable.

In the summary of information I received, the two SAT9 math scores for both the district and Alta Mesa were lumped into the title Total Math. The district showed a percentile score of 43; Alta Mesa 44; Alta Mesa's second grade 45. The percentile score for my own class was 50—nothing to shout about, but encouraging. For my own class, I had the information to do a break-out of the

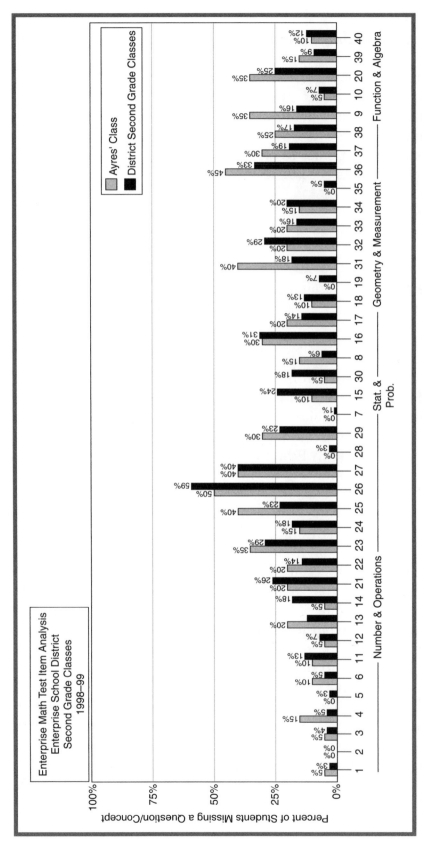

Figure 6.6. Math test item analysis comparing Mrs. Ayres' class with the district second grade classes. Used with permission of the Enterprise School District.

two scores—procedures and problem solving. Since problem solving had been the thrust of my learning experiment, I was interested in looking at these numbers. The class as a whole scored 56th percentile in problem solving. I still wasn't shouting, but I was smiling. I wondered about my seven math club students, the ones who had continually pressed for more. This group was made up of four girls and three boys. Three of the seven were ESL students (English is their second language). As a group, they scored 80th percentile in problem solving! I was about to shout, but I realized that was cheating. They were my top seven students. The realization did nothing to wipe the smile from my face. A further look showed me that my ESL students actually scored better in procedures than in problem solving, undoubtedly a result of the language barrier.

I decided to look at the scores of my After School Club, the ones who needed extra support, and were willing to work hard to improve. Their scores were much lower, but what was the highest SAT9 score for each child, looking at all subjects tested? Problem solving! Now there's something to shout about.

An axiom of truth for my own life has been "What you focus upon, grows." In looking at the SAT9 scores, I could see the results of my intense focus on problem solving. My choice of focus, which appeared at first to be a tight little ball of problems, opened up to reveal solutions, when viewed in the light of the line graph. I saw this ball become a single line, as it unwound and crawled across the grid paper. I have not mentioned that I also collected data, and tracked growth in spelling throughout the year. In this area, my class SAT9 percentile score was higher than that of both Alta Mesa's and the Enterprise District's second grade. From looking at these numbers, I now realize that the areas of study I choose to track are the ones that will show the most growth.

Pondering the Meaning of What I Learned

My mind started spinning through all the numbers. Could my interpretation of the numbers give me the information I needed to decide if I should stay on my present course, or get out and lay some new track? Were the numbers the only thing I should consider? I thought of a couple of reasons to broaden my thinking past just the numbers in making decisions about the future. My experience throughout the years I have taught is that each year the individuals that make up my class create a unique blend, different from any other. Sometimes more students are at the bottom struggling toward understanding; sometimes more at the top. Using more than one year of data is needed to strengthen the case that numbers present.

A second reason for using more than just numbers relates to poverty level. Before coming to Alta Mesa, I taught second grade at a school that had the highest SAT9 scores in the county last year. The fact that it is also one of the highest socioeconomic areas is no coincidence. If I went strictly by numbers, I would have to throw out this year's learning experiment as a total failure. My numbers at the other school were much better. No; numbers cannot be the only consideration.

Continuous Improvement

Numbers coupled with the concept of continuous improvement is the bottom line. I use data to guide me in creating continuous improvement, taking children from where they are now to desired new heights of knowledge and understanding. Benchmarks are my guide; PDSA is my vehicle. My ultimate goal is to bring about improvement in student learning. To do this, I continually look for ways to improve what I am doing. If I had one class I could take through from kindergarten to eighth grade, I could see the effects of my teaching over time. But I don't need to do that. Instead of one ant dragging the crumb all the way to the nest, a team of ants gets it there. The concept of continuous improvement makes the most sense when top management supports schools in developing a continuum as a ladder for children to climb up through the grade levels; and supports teachers in working together to figure out how to do it. All teachers desire to send their students out the door at the end of the school year a few rungs higher than the year before. Each teacher is more able to do this if the teacher in the grade just below has also given an extra boost. Continuous improvement is about standing on the shoulders of those who have gone before. Classroom data is a powerful feedback tool that supports adherence to the continuum. This, along with the year-end assessments, is used to create a picture that guides me toward my goal of continued improvement.

Maybe the idea of continuous improvement is as silly as having everyone get above the 50th percentile. Does continual improvement go on into infinity? Does the human potential have a cap? People used to think there was a four-minute cap on how fast someone could run a mile. Then Roger Bannister ran it under four minutes. Since then, many people have run a less than four-minute mile. It bears out the philosophy, "What we can believe, we can achieve." With the evolutionary push of life to organize and then reorganize into yet higher and higher levels of understanding, I would like to think there is no cap on human potential. In education, the possibilities for testing this theory are wide open.

When I am discouraged as I work with a struggling child, I focus on any improvement I can see, however small. At the same time, my desire is to increase the rate of growth. Because it is in children's nature to want to learn, I can show up every day and do almost anything in the way of teaching, and they are going to learn. But the time factor cannot be ignored. The cactus my friend gave me three years ago has grown about three inches. When my son and I measured the growth of a new bamboo shoot, in a 24 hour period of time, it grew 13 inches! As with my plants, in my classroom, I am measuring growth against time. I want the rate of my class' continuous improvement to be more than that of the cactus. And, of course, the next year I'd be ecstatic if the rate of improvement increased from the year before as a result of something I figured out to do differently. That is what the PDSA cycle is all about.

The Value of Data Collection Coupled with the PDSA Cycle

With these thoughts, I came face to face with my question: What was there about my data collection, coupled with the PDSA cycle, that would have me continue a similar plan into the next school year? It is not until a question is asked that an answer comes. I saw that the real value of my learning experiment was in the clarity it provided for me about where I was going. When I placed that clarity upon large wall charts, I was asking all who looked at them to keep me accountable. In past years I have worked hard, but I have been less likely to notice if something was not working. I would look at benchmarks every now and then, whenever I found them buried under bunches of other papers, where no one could support me with them. When the end of the year came, I either felt good or was disappointed with individual student or class performance, but it was not easy to connect the performance with my teaching. In problem solving this year, I know exactly what I did to improve performance. I turned the steering wheel, and the car started going North on the map. *The real value of my learning experiment was in the clarity that was provided by my road maps, along with the level of accountability the road maps inspired, first in myself, and then in my students and their families.* When I state to the world what I am about, and I openly invite others to support me in getting there, the ball does not get dropped. And instead of being carried by one, it is carried by many.

As I am finishing this book, it is fall, and school has begun once again. On the playground, Nick came running up to me. "Mrs. Ayres, I wish I was in your class again." I said, "I wish you were in my class for the rest of my life." Nick's face broke into a big grin, and he said, "I know why. It's because I'm so good in math." I wondered if anyone had talked to him about his SAT9 scores. I found no one had. He doesn't know he scored in the 99th percentile on problem solving. It doesn't matter. He knows from the inside out that he is good at something he loves. His self-esteem is oozing from every pore of his body. I was supported in getting him there by tussling with the three math graphs in my room—the Enterprise Weekly, Problem Solving, and Math

Enthusiasm. In looking out the train window, I've decided it is okay, for this next school year, to end up where I am headed.

Summary

In this chapter, I looked at the value of reflecting daily on what worked in the classroom and what didn't. I expanded the reflection backwards to view the entire school year. I included students in the reflective process by teaching them to use a fishbone diagram. Also, how students ended up feeling about mathematics was important to notice. I used the Pareto chart for a reflecting tool, discovering that most of my students' weaknesses with math information centered around place value, measurement, and geometry. This gave me the clay to mold for next year's planning. I examined the value of using quality tools for thinking, and realized they give perspective and clarity to what otherwise seems muddled and complicated. Using results data provided to me by the district, I compared my scores to those of others in my school, as well as those of the district as a whole. With the SAT9 scores, numbers were put together and taken apart in different ways to give me a picture of my year. I concluded that, in any school year, regardless of where my students begin in the fall, continuous improvement is the desired outcome. Working with teachers at the other grade levels as well as my own, to accomplish this goal, brings the most widespread benefit. Using process data and benchmarks, that all are invited to view and support, keeps me accountable to these goals.

As I finish putting this book together, I am thinking about the direction the Enterprise School District will be taking in the next few years. This past year, through reading, and taking some district workshops in the fall, I began to form a picture of Dr. Jenkins' vision of a working system. Bringing about such a miracle requires a clear focus on the possibility. Nothing can come about that has not been first created in thought. As top management, Dr. Jenkins was keeper of the thought. Because he was recruited to be superintendent in another school district mid-year, I'm now wondering what will happen in the Enterprise school system. Dr. Jenkins' efforts have moved the Enterprise district toward the ideal that Dr. Deming established, but much more education is needed. A change of thinking might come about in a short period of time, but implementing it does not. Without management's vision and commitment for a whole, working system, we can still be an excellent school district with lots of good people trying their best, but we will continue to be like hundreds of other school districts with lots of good people trying their best. Without the model, our great potential is swallowed up in working harder, not smarter.

I think about the beauty of the hologram and the metaphor it is for that potential. When a rock hits the clear, smooth water of a lake, the concentric rings that pulse out are a pattern of circles spreading, to the shorelines. The people in the system are the rocks that begin the rings. As one's waves touches another's, an interference pattern is established. With these interconnecting waves one can view any part, and see a picture of the whole.

Notes

1. Barbara A. Cleary and Sally J. Duncan, *Tools and Techniques to Inspire Classroom Learning* (ASQ Quality Press: Milwaukee, WI, 1997), 57.
2. Elaine McClanahan and Carolyn Wicks, *Future Force, Kids That Want To, Can, and Do! A Teacher's Handbook For Using TQM in the Classroom* (PACT Publishing: Chino Hills, CA, 1993), 155.

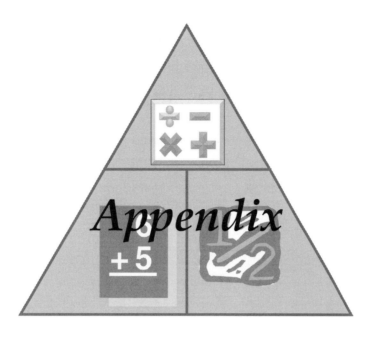

Edward Deming's Fourteen Points

1. Create constancy of purpose.
2. Adopt the new philosophy.
3. Cease dependence on mass inspection to achieve quality.
4. End the practice of awarding business on price tag alone. Instead, minimize total cost, often accomplished by working with a single supplier.
5. Improve constantly the system of production and service.
6. Institute training on the job.
7. Institute leadership.
8. Drive out fear.
9. Break down barriers between departments.
10. Eliminate slogans, exhortations, and numerical targets.
11. Eliminate work standards (quotas) and management by objective.
12. Remove barriers that rob workers, engineers, and managers of their right to pride of workmanship.
13. Institute a vigorous program of education and self-improvement.
14. Put everyone in the company to work to accomplish the transformation.

Notes

Lloyd Dobyns and Clare Crawford-Mason, *Quality or Else: The Revolution in World Business* (Houghton Mifflin Company: New York, 1991), 289.

Name _____

MATH ENTHUSIASM

Mark one choice.

I hate math.	
I don't like math.	
Math is okay.	
I like math.	
I love math.	

Write about your choice.

_____

_____

_____

_____

_____

_____

_____

_____

_____

_____

HUNGRY BUG ADDITION

PREREQUISITE: Experience with number combinations to ten (mastery not needed) and progress to 100 on the counting strip.

MATERIALS: Base Ten Blocks.

To solve   9   Take out nine units and line them up beside a long:
             + 4

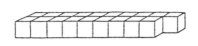

Nine is a "Hungry Bug" that wants to be ten; how many more does he need? (One) If he gobbles up one of the four, what is left? (3) The sum is 10 + 3.

What happens to the Hungry Nine Bug? Use a long to measure and see:

9	9	9	9	9	9	9	9
+ 2	+ 5	+ 7	+ 3	+ 9	+ 6	+ 4	+ 8

What is the rule?

Eight is a Hungry Bug with an appetite for how much? (Yes, two.)

To solve: 8   Take out eight units and line them up beside a long:
             + 3

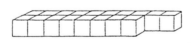

Eight is a "Hungry Bug" that wants to be ten; how many more does he need? (Two) If he gobbles up two of the three, how many are left?

8	8	8	8	8	8	8	8
+ 3	+ 5	+ 7	+ 9	+ 2	+ 4	+ 6	+ 8

Seven is a Hungry Number Bug that is even hungrier. Can you find out about 7?

Source: Exercises on pages 147–152 courtesy of Peggy McLean, Mary Laycock, Elsie Robertson, *Skateboard Practice, Addition, Subtraction.* © 1990 Activity Resources Co., Inc., Hayward, CA. Used with permission.

SPIN A FLAT

Game for four—three players and a banker

PREREQUISITE: Experience with the counting strip; sums to ten, not mastered.

MATERIALS: Base ten blocks, die, and computer mats for three children.

PROCEDURE: One child is banker. Each child spins the die to find the highest, who goes first. Play continues clockwise. On each spin, the child asks the banker for the number of units indicated and places them on the units space of the computer. When there are ten, the child matches and asks the banker for a long. Whoever trades ten longs for a flat first wins.

MY COMPUTER			
blocks	flats	longs	units

GO BROKE

PREREQUISITE: Experience of playing Spin a Flat.

MATERIALS: Base ten blocks, die, three computer mats.

PROCEDURE: Each player begins with a flat. One child is banker; the other three spin the die to see who spins the lowest. The lowest goes first; play proceeds clockwise. Each child spins, the number spun must be removed. The banker will have to be asked for change so the removal can be accomplished. The winner is the first to reach zero exactly. (The number of the last spin must be the exact number of blocks remaining.)

Names for 8

Use mat and cubes
to find names for 8.

Prove with rods and crayons.

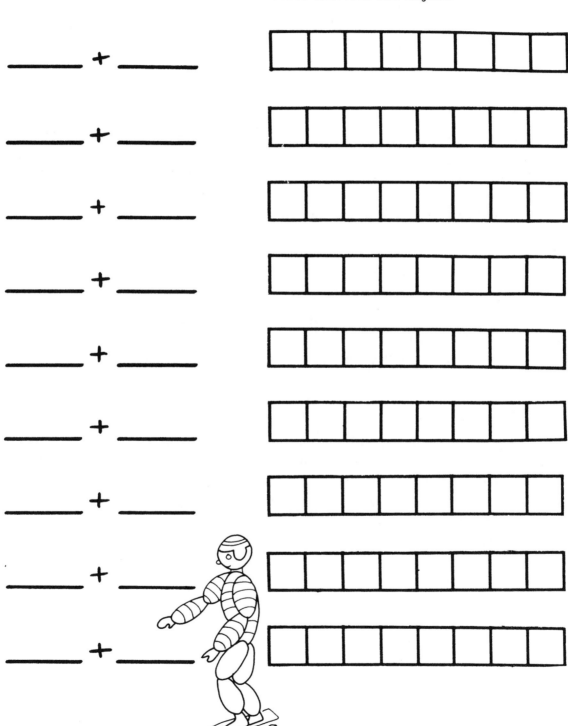

_____ + _____

_____ + _____

_____ + _____

_____ + _____

_____ + _____

_____ + _____

_____ + _____

_____ + _____

_____ + _____

149

SHOOTING FOR 10's

You need:
Two players
"Shooting For 10's" game mat for each player
Cuisenaire rods or the paper facsimile
Crayons to match the color of the rods

Game 1.
Purpose: To associate a number value to the rod colors.

Children toss to see who gets the higher number. That child goes first. The players toss the die, choose the rod that matches the number thrown, and color the same number of squares on the game mat that match the color of the rod. Winner is the first to exactly color the entire grid. A child may be ahead but lose several turns before tossing the number that is needed. Rods can be colored horizontally and vertically.

Game 2.
Purpose: To practice sums less than 12.

The game is the same except that two dice are used and the child may choose one rod that is the sum, if possible, or may use rods that represent the separate dice. For example, if two sixes are thrown, ONLY two dark green rods can be used. As the board becomes full a child may use only one die. If two dice are used ALL must be colored or the turn is lost!

Game 3.
Purpose: To practice writing names for ten.

The game is the same except that all rods must be colored horizontally. The score is the number of rows of ten that can be filled and the appropriate number sentence written. For example, if one throw is 3 and 3, a dark green six, or two light green threes may be colored. If the next throw is a 4 and 5, the child would color the purple four on the first row and color the yellow five on the next row. Either $6 + 4 = 10$ or $3 + 3 + 4 = 10$ would be written on the line after the colored row. Each correct number sentence for ten counts one point. Winner is the child with the most points.

Shooter:

Shooting for 10's

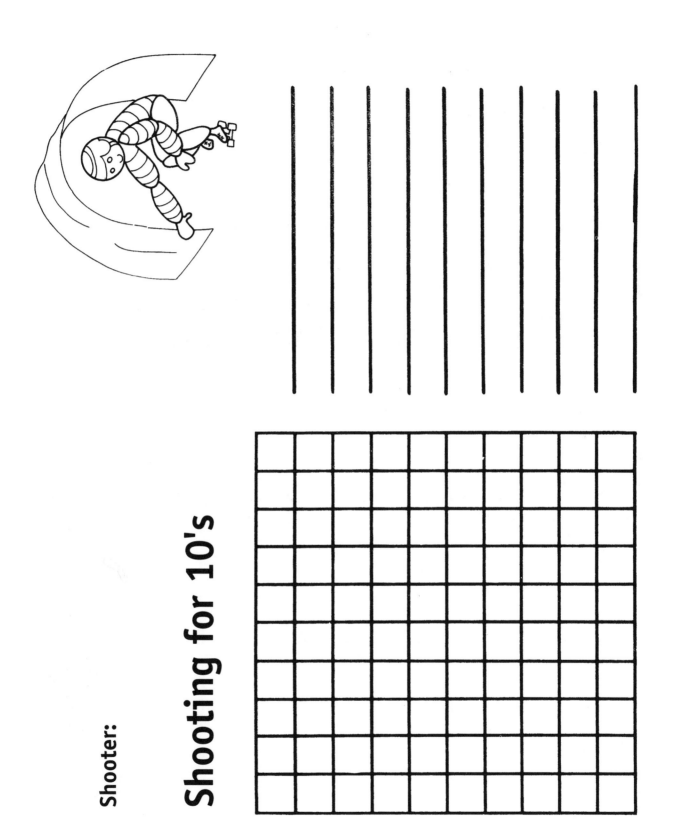

TEN SPOTS

1. Get a set of double-nine or double-ten dominoes.

2. Ask one or two friends to play.

3. Take 5 dominoes each.

4. Decide who goes first.

5. Take turns putting down dominoes so that when they touch, they make 10 spots.

6. Take a new domino each time you have a turn.

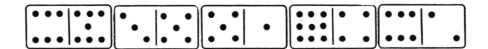

If you have a domino with 12 spots, put it with one that has 2 spots, because 12 − 2 = 10.

Hints:

If you have a domino with 0 spots, put it with one that has 10 spots, because 10 + 0 = 10.

THE COUNTING STRIP

Have the student set up their Place Value Board, Base Ten Blocks and Counting Strip like this.

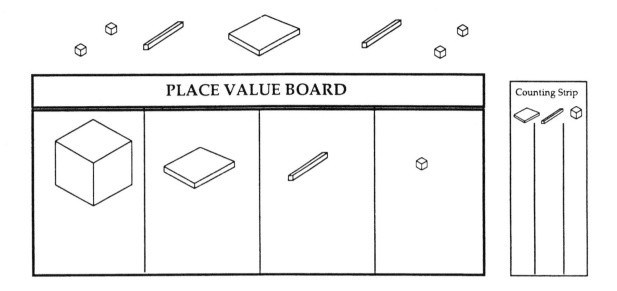

CAUTION
Do the activity with 2 or 3 students.

Procedure:

The student places a unit on the Place Value Board in the units place and records 1 on the counting strip. He/She adds another unit and records 2, another unit and records 3, etc. When ten units are on the Place Value Board, the student matches the ten units to the long and now writes 1 in the longs place and 0 in the units place. 10 is verbalized as one long and no units.

As each ten is completed the student matches 10 units to make another long. This should continue until the student is in the thirties or forties or until the alloted time for the lesson ends. When it is time for the next lesson, the student takes out the Place Value Board and continues counting from the number recorded last.

To maintain the motivation to complete the strip to 1,000, a certificate should be awarded.

Practicing Place Value pages ask the student to identify pictures of blocks. Some students may need to build the pictures before answering. Build and stamp more examples as needed.

			1
			2
			3
			4
			5
			6
			7
			8
			9
		1	0
		1	1
			.
			.
Paste			

Source: Exercises on pages 153–154 courtesy of Peggy McClean; Mary Laycock, Margaret A. Smart, *Building Understanding with Base Ten Blocks*. © 1990 Activity Resources Co., Inc., Hayward, CA. Used with permission.

MAKE A COUNTING STRIP

Cut the strips apart and paste together to make one long strip.

PASTE			

PASTE			

PASTE			

THE DOUBLES

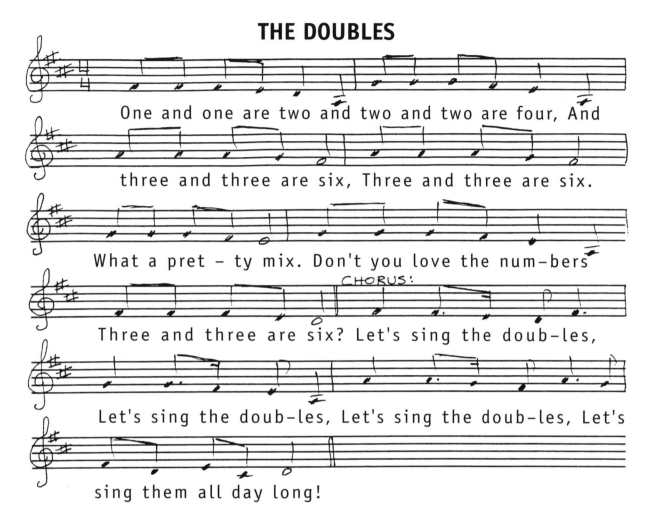

One and one are two and two and two are four, And

three and three are six, Three and three are six.

What a pret - ty mix. Don't you love the num-bers

CHORUS:

Three and three are six? Let's sing the doub-les,

Let's sing the doub-les, Let's sing the doub-les, Let's

sing them all day long!

Four and four are eight, and five and five are ten,
And six and six are twelve, six and six are twelve,
Say it to yourselves.
Don't you love the numbers six and six are twelve?

Seven and seven are fourteen, eight and eight are sixteen,
Nine and nine are eighteen, nine and nine are eighteen,
Yes, they make a great team,
Don't you love the numbers nine and nine are eighteen?

Source: Created by Carolyn Ayres.

THE DOUBLES

One and one are two and two and two are four,
And three and three are six. Three and three are six.
What a pretty mix. Don't you love the numbers
Three and three are six?

Chorus:
Let's sing the doubles, Let's sing the doubles,
Let's sing the doubles, Let's sing them all day long!

Four and four are eight and five and five are ten,
And six and six are twelve, six and six are twelve,
Say it to yourselves.
Don't you love the numbers
Six and six are twelve?

Seven and seven are fourteen, eight and eight are sixteen,
Nine and nine are eighteen, nine and nine are eighteen,
Yes, they make a great team.
Don't you love the numbers
Nine and nine are eighteen?

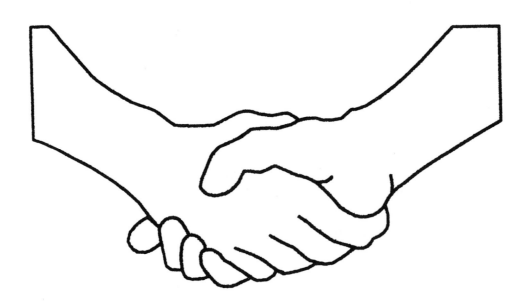

RON JOHNSON'S FRACTION HUNT

1. The first half of go + the last ½ of go.
2. The opposite of under.
3. The last $^{2}/_{7}$ of Grigsby.
4. The XX letter of the alphabet + the first ⅓ of helium.
5. The first $^{4}/_{9}$ of backwoods.
6. Do + the XV letter of the alphabet + the first letter of the word that describes the distance from the center of a circle to its edge.
7. The last ¾ of sand.
8. The first ⅙ of dodecahedron (a 12-sided polygon).
9. The answer to $3 + 5 \times 6 - 23$.
10. The opposite of pull.
11. The big brown truck that delivers packages.
12. The first ⅞ of proceeds.
13. The answer to 4 squared minus 14.
14. The first ½ of Thelma.
15. The XX – V letter of the alphabet + a + the first ¼ of kite.
16. The first $^{3}/_{7}$ of treason + the last ½ of bees.
17. &
18. The first ⅗ of there.
19. The opposite of big.
20. The first ⅜ of cupboard.

Source: Exercises on pages 157–182 courtesy of Ron Johnson. Used with permission.

FRACTION HUNT CRYPTIC CODE AND FURTHER CLUES

Problem: Go over by the back door and do 10 push ups. Proceed to the oak trees and the little cup.

Jog over to the middle backstop. Stand facing Swasey to the west. Turn about 60 degrees counterclockwise. Proceed about 30 meters ahead to the clue in red.

That was fun, but I'm sorry to say.
You have to solve this problem to win the day.
Check in the hanging file of student # 2 + 3 $\times$ 6 − 7.

You're real close to winning the prize,
The sum of the angles in a triangle is before your eyes.

ESTIMATION STATIONS

Integrating Mathematics
and
Physical Fitness

ESTIMATION STATIONS
Integrating Mathematics and Physical Fitness

Math Skills

- Estimation
- Calculator use
- Measurement practice using metric units
- Addition and subtraction skills
- Rounding different types of numbers
- Averaging data
- Mean/mode/median
- Graphing data
- Fraction/decimal/percent practice

Fitness Benefits

- Work on different components of fitness
  - cardiovascular/heart & lungs
  - flexibility
  - strength
  - balance
  - sports skills/hand-eye coordination
- Improve self-awareness
- Improve social development and social interaction
  - cooperate with a small group
  - taking turns
  - accept and respect performance of others regardless of ability level
  - using low voices (the 12" voice)
  - staying with an assigned group

Task

Today you are going to work in teams to estimate and measure. First you will estimate various measures of distance and time. Then you will perform each event and actually measure the distance or time. Finally, you will find the difference between your estimates and the actual measurements. All scores will be recorded on your worksheets.

Procedure

- Teacher demonstrates each station. (don't show too much)
- Students write their estimates after each demo. (write in pen)
- Explain rotation of groups. Use a chart on the board.
- Explain roles for each group member. (wall chart)
- Explain how you will assess students.
- Assign each group to a station. Practice rotation.
- Captain reads instructions to team.
- Each team member performs the activity and records their measurements on their score sheets.
- After 3–5 minutes, ask "Anyone need extra time?" Signal all groups to rotate.
- When teams have completed all activities, they are seated.
- The score is the difference between their estimate and their actual measurement. Calculate total score.
- Team members exchange and check each other's papers.
- Calculate team total score.
- Pick out something positive to say about each team.
- Winning team is the team with the lowest total score.
- Prizes awarded to winning team and individuals. (optional)
- Teacher may discuss how a low score shows accuracy in estimating. This activity not dependent on being a great athlete.

Roles

Captain: Reads station directions to group. Makes sure that all members of the team complete each activity and the group rotates correctly through all stations.

Noise Monitor: Reminds group to maintain a low voice.

Judge: Makes sure that all members are accurate and correctly recorded.

Checker: Makes sure that all group members leave their station neat and orderly, with all materials in place.

Student volunteers may be used as official timers. They will not compete.

Student Self-Assessment

Discuss with your team the following questions and come to an agreement on the best answer.

1. Did we:

   a) Take turns   _____ yes   _____ no

   b) Use 12" voices   _____ yes   _____ no

   c) Stick to our roles   _____ yes   _____ no

   d) Stay in our group   _____ yes   _____ no

2. Write some of the things we said/did that helped or hindered our team work. What can we do differently next time to make it better?

   a) Taking turns

   b) Using 12" voices

   c) Sticking to your role

   d) Staying in your group

3. When you are done, and your team understands and agrees with the above answers, sign your names below and turn this sheet in with your worksheets.

   _____   _____

   Signature   Signature

   _____   _____

   Signature   Signature

TEACHER OBSERVATION SHEET

Observe each group at least 2 times for the same amount of time. Measure the frequency with which each behavior occurs by recording tally marks.

Group Number	Use 12" Voices	Everyone Gets a Turn	Stay with Their Group	Use Materials Appropriately	Comments
1					
2					
3					
4					
5					
6					
7					
8					

ESTIMATION STATIONS

Name _____

Team # _____

Event	Estimate	Actual	Score (Difference)
Sponge Throw	m.	m.	
Standing Long Jump	cm.	cm.	
Jump & Mark	cm.	cm.	
Fast Hands	cm.	cm.	
Five-Meter Eraser Balance	sec.	sec.	
Jump Rope	# rep.	# rep.	
Five-Meter, One Leg Dash	sec.	sec.	
Pulse Rate After Exercise	bpm.	bpm.	
		Total	

*Round all scores to the nearest whole number.

ESTIMATION STATIONS

Name _____

Team # _____

Event	Estimate	Actual	Score (Difference)
		Total	

*Round all scores to the nearest whole number.

Team Score Sheet

Team # ➡️

Name	Score
Total Team Score	

CHECKER

Makes sure that all group members leave their station neat and orderly, with all materials in place.

JUDGE

Makes sure that all members have recorded accurately.

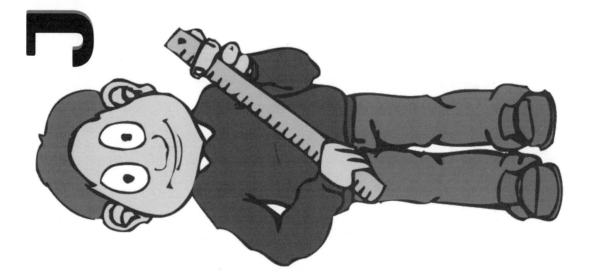

NOISE MONITOR

Reminds group to maintain a low voice.

CAPTAIN

1. Reads station directions to group.
2. Makes sure that all members of the team complete each activity.
3. Makes sure the group rotates correctly through all stations.

Straight Arm Dumbbells

Hold a dumbbell in each hand and put your arms straight out to the side of your body. Hold as long as you can. Record the number of seconds.

Fast Hands

Reaction Time

A meter stick is held vertically directly above your outstretched hand. The stick is dropped. Grab the stick with your hand as soon as you see it start to drop. Record the number of centimeters on the stick even with the top of your hand.

Sponge Toss

Place feet on starting line. Throw sponge. Measure the distance from starting line to position of the sponge. Take 2 tosses, the longest counts. Record in meters.

Pulse Rate After Exercise

On signal, all team members run in place for 30 seconds. After running, take pulse by placing two fingers on neck or wrist. On signal, count pulse beats for 30 seconds. Multiply by 2. Record pulse rate.

Standing Long Jump

Place both feet on starting line. Jump forward. Measure distance from starting line to heel of foot closest to starting line. Record best of two jumps in centimeters.

Jump & Mark

Stand next to paper. Hold marker. Jump and reach as high as you can and make a mark on the paper. Take two turns, highest mark counts. Record in centimeters.

5 Meter Eraser Balance

Stand behind starting line.

Balance eraser on your head.

On signal to start, walk as quickly as possible to the finish line. Record # of seconds on your scorecard.

Jump Rope

On signal to start, jump rope as quickly as possible for 15 seconds. Count number of times you jumped during that time. Record number of jumps completed on your scorecard.

5 Meter One-Footed Dash

Stand behind starting line. Balance on one foot. On signal to start, hop on one foot until you reach the finish line. Record number of seconds on your scorecard.

INDOOR STATION IDEAS

Cardiovascular Heart & Lungs

jump rope
jog in place
shuttle sprints
pacer
line jumps
paper bag jump
hula hoop
one leg hops
jumping jacks
coordinated jacks
scissor jacks
kick out—touch toes
knee clap (clap under leg)
kick out & touch toes
toe touches
under bridge and around
side steps

Strength

push-ups
sit-ups
chin-ups
low bar pull-ups
squats
straight arm weights
chair push-ups
spring machine
flexed arm hang
chair dips
lame dog (left leg off ground)
chair steps
stand/sit/stand/sit
flutter kick
5 meter crab walk
leg lifts & hold above 40 cm.
hand walk (push-up position)
foot launcher
leap the brook

Flexibility

sit & reach
limbo
stretches
under over bars

Balance

line walks
2 × 4 walks
beam on milk cartons
hop with Barney on head
heel/toe line walk
log walk
5 meter stilts
cone and cord maze (crawl through)
tip toe 5 meter line
foot twister
one leg through styrofoam cup
course
limbo

Sports Skills/Hand-Eye

dribble through cones
dribbles in 30 seconds
scarf juggling
shoot baskets
badminton hits
nerf throw
Frisbee throw
catch bean bag behind back
bowling (styrofoam cups taped & tennis ball)
tire jump
shot-put Nerf
waste basket target toss (5, 10, & 20 pts)
scoop and fluff ball
darts
stack blocks
chair ring toss
keep it up (birdie/racquetball)
blindfold and catch (Billy Balls)

students design stations
playground equipment stations

P.E. MATH STATIONS

Ron Johnson's P.E. Stations Adapted to Use with Grades 2–4

Balancing on One Foot
Estimate how many seconds you can think you can stand on one foot.
Have a partner time you. Balance on one foot for as long as you can. Take your longest time.
Figure the difference between your estimate and the actual time.
Bounce a ball the number of times that is the difference.

Sit-Ups
Estimate how many sit-ups you think you can do in one minute.
Get a partner to time you. Count the number of sit-ups you can do in one minute.
Figure the difference between your estimate and the actual count.
Touch your toes the number of times that is the difference.

Standing Broad Jump
Estimate the number of inches you think you will jump.
Do the standing broad jump. Get someone to mark the position of your heel each time.
Measure your jump in inches. Figure the difference between your estimation and your actual jump.
Do jumping jacks the number of times that is the difference.

Jar Estimation
Estimate the number of things in the jar.
Figure the difference between your estimation and the actual amount. (Answer is taped to the bottom of the jar.)
Jump a rope the number of times that is the difference.

Balance Beam
Estimate how many heel-to-toe steps it will take you to get from one end of the balance beam to the other.
Figure the difference between your estimation and the actual count.
Frog hop the number of times that is the difference.

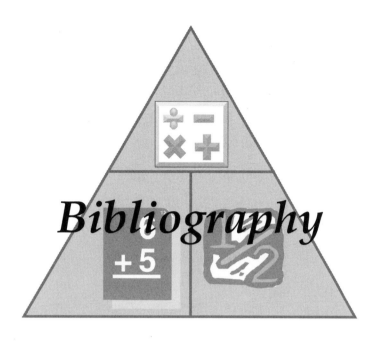

Connecting Mathematics to Literature

Aker, Suzanne. *What Comes in 2's, 3's, & 4's?*. New York: Scholastic Inc., 1990.

Anno, Mitsumasa. *Anno's Counting Book*. New York: Scholastic Inc., 1975.

Giganti, Paul. Jr., *Each Orange Had 8 Slices*. New York: Trumpet Club, 1992.

Gretz, Susanna. *Teddy Bears 1 to 10*. New York: Macmillan Publishing Company, 1986.

Grossman, Bill. *Tommy at the Grocery Store*. New York: HarperCollins Publishers, 1989.

Hoban, Tana. *26 Letters and 99 Cents*. New York: William Morrow and Company, Inc., 1987.

Holtzman, Caren. *A Quarter from the Tooth Fairy*. New York: Scholastic Inc., 1995.

Hooks, William H. *A Dozen Dizzy Dogs*. New York: Bantam Doubleday Dell Publishing Group, Inc., 1990.

Hutchins, Pat. *The Doorbell Rang*. New York: Scholastic Inc., 1986.

Meltzer Kleinhenz, Sydnie. *More For Me!* New York: Scholastic Inc., 1997.

MacDonald, Suse, and Bill Oakes. *Puzzlers*. New York: Scholastic Inc., 1989.

Mathews, Louise. *Bunches and Bunches of Bunnies*. New York: Scholastic Inc., 1978.

McMillan, Bruce. *Eating Fractions*. New York: Scholastic Inc., 1991.

Munsch, Robert. *Something Good*. New York: Annick Press Ltd., 1990.

Myller, Rolf. *How Big is a Foot?* New York: Bantam Doubleday Dell Publishing Group, Inc., 1962.

Pinczes, Elinor J. *One Hundred Hungry Ants*. New York: Scholastic Inc., 1993.

Pinczes, Elinor J. *A Remainder of One*. New York: Scholastic Inc., 1995.

Pluckrose, Henry. *Math Counts: Capacity*. Chicago: Childrens Press, 1995.

Rocklin, Joanne. *How Much Is That Guinea Pig in the Window?* New York: Scholastic Inc., 1995.

Sandoval, Dolores. *Be Patient, Abdul*. New York: Simon & Shuster Children's Publishing Division, 1996.

Schlein, Miriam. *More Than One*. New York: Scholastic Inc., 1997.

Schwartz, David M. *How Much Is A Million?* New York: Scholastic Inc., 1985.

Schwartz, David M. *If You Made A Million*. New York: Scholastic Inc., 1989.

Sheppard, Jeff. *The Right Number of Elephants*. New York: Scholastic Inc., 1991.

Slater, Teddy. *Just a Minute!* New York: Scholastic Inc., 1996.

Tankel, Lara. *What's the Time?* New York: DK Publishing Inc., 1995.

Viorst, Judith. *Alexander, Who Used to Be Rich Last Sunday.* New York: Simon & Shuster Children's Publishing Division, 1978.

Wheatley, Nadia. *1 is for One.* New York: Mondo Publishing, 1996.

Williams, Vera B. *A Chair For My Mother.* New York: William Morrow & Company, Inc., 1982.

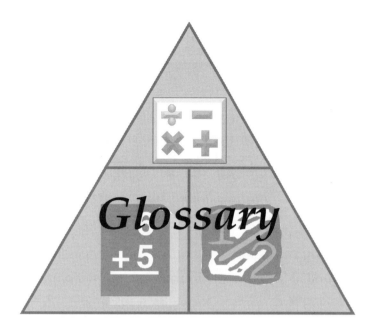

Aim a concise statement of purpose.

Algorithm any particular procedure for solving a certain type of problem, as the rule for finding the greatest common denominator.

ASQ a society of individuals and organizational members dedicated to the ongoing development, advancement, and promotion of quality concepts, principles, and technologies.

Class Action software developed that allows the educator to create computer-generated quality measurement tools for monitoring student and class growth. It includes scatter diagram, student run and class run charts, scatter diagram with overlay of student run chart, histogram, Pareto chart, cause and effect diagram (fishbone), and web chart.

Cause-and-effect diagram a tool for individual or group problem solving that provides a way to generate and categorize causes for a given effect. Also known as a fishbone diagram or Ishikawa diagram.

Common cause variation causes that are inherent in any process all the time. A process that has only common causes of variation is said to be stable or predictable.

Continuous process improvement includes the actions taken throughout an organization to increase the effectiveness and efficiency of activities and processes in order to provide added benefits to the customer and organization. It is considered a subset of total quality management and operates according to the premise that organizations can always make improvements. Continuous improvement can also be equated with reducing process variation.

Deming, W. Edward (deceased) a prominent consultant, teacher, and author on the subject of quality. After he had shared his expertise in statistical quality control to help the U.S. war effort during World War II, the War Department sent Deming to Japan in 1946 to help that nation recover from its wartime losses. Deming has published more than 200 works, including the well-known books *Quality, Productivity, and Competitive Position* and *Out of the Crisis.* Deming, who developed the 14 points for managing, is an ASQ Honorary member.

Enterprise Weekly   a five-question weekly test created by math consultants in the Enterprise School District. Questions are randomly chosen from a bank of questions created for each grade level. The Enterprise Weekly tests students mastery of information.

Epistemology   the study or theory of the origin, nature, methods, and limits of knowledge.

Feedback   comments on success or failure of a particular program from customers.

Fishbone diagram   (cause and effect diagram) a diagram that illustrates causes and subcauses that lead to an effect. It looks like a fish skeleton.

Flowchart   a graphic portrayal of a process, showing the steps that are involved in that process and their relationship to one another.

Histogram   a graphic summary of variation in a set of data. The pictorial nature of the histogram lets people see patterns that are difficult to see in a simple table of numbers.

Hologram   a negative produced by exposing a plate near a subject illuminated by a laser. When placed in a laser beam, a true three-dimensional image of the subject is formed. Unlike normal photographs, every portion of a piece of holographic film contains all of the information of the whole.

Information   facts about the past. Facts are tools used for thinking mathematically.

Knowledge   the ability to use information to create a better future by relating the past to present and future situations through problem solving.

KWL chart   before a new concept or unit is introduced students are asked to list things they already **Know** about the subject. Next they list what they **Want to learn.** This gives the teacher a place to begin with the unit. After the unit is completed, students list what they **Learned.** Often the three areas are listed side by side on one chart.

Long another word for a group of ten, sometimes called a ten stick. The word "long" can be used to describe the block that is created when the first trade takes place from units to a connected group, when counting by ones. "Long" can be used in any base, whereas "ten stick" can only be used in base ten. Along this same line, a hundred block is a flat, and a thousand block is a cube.

Mentors those who have been through a given program or experience, who help newcomers to the program.

Pareto chart a basic tool used to graphically rank causes from most significant to least significant. It utilizes a vertical bar graph in which the bar height reflects the frequency or impact of causes.

PDSA cycle plan-do-study-act cycle. A four-step process for quality improvement developed by statistician Walter Shewhart and emphasized by W. Edward Deming. A different way of looking at the scientific method to solve systemic problems in the classroom.

Plus-delta chart a way to get feedback from students. It is a simple chart with two columns. One side is the plus side where students list what they like; the other side is the delta side where they list the things that need to change.

Process the students and teacher work together as a team in the classroom to refine and improve the incremental steps in which the concepts are learned and applied.

Process data data collected each week.

Results data data collected at the end of the school year.

Run chart a form of trend analysis that uses a graph to show process measurement on the vertical axis against time. In this book, it is a graph used to record test results on which the number correct is the vertical axis and the particular week is the horizontal axis. From this information trends and patterns can be observed. A student run chart shows an individual student's progress. A class run chart combines individual student scores to show how the class as a whole is doing.

Scatter diagram a graphical technique to analyze the relationship between two variables. Two sets of data are plotted on a graph, with the y axis being used for the variable to be predicted and the x axis being used for the variable to make the prediction. The graph will show possible relationships (although two variables might appear to be related, they might not be: those who know most about the variables must make that evaluation). It is a statistical tool that records each student score as a dot. This personalizes what otherwise would be a total score lumped as one number (as with the class run chart).

Special cause variation causes of variation that arise because of special circumstances. They are not an inherent part of a process. Special causes are also referred to as assignable causes. In the classroom, this would be an event outside the normal classroom environment that affects the steady improvement of the class, or of the individual student, as indicated by the run charts. Variation that is created by factors that are universal to the system is considered common causes.

Standard a statement, specification, or quantity of material against which measured outputs from a process may be judged as acceptable or unacceptable.

Supply the upcoming or brand new students that start the year in each room.

System a network of connecting processes which work together to accomplish the aim of the system. The parts of a system are aim, supply, input, process, output, customers, and feedback. In education, each part is further defined so the terms apply specifically to that system.

Total Quality Management (TQM) TQM utilizes statistical and problem-solving tools to bring about planned change and continuous improvement in a system.

Variation a common characteristic of systems. Variation can be analyzed by means of appropriate statistical tools so that it can be reduced and improvement can ensue. Variation may be due to common causes or special causes.

Notes

Glossary descriptions are adapted from the following sources.

Barbara A. Cleary and Sally J. Duncan, *Tools and Techniques to Inspire Classroom Learning* (ASQ Quality Press: Milwaukee, WI, 1997).

Jeffrey J. Burgard, *Continuous Improvement in the Science Classroom* (ASQ Quality Press: Milwaukee, WI, 2000).

Karen R. Fauss, *Continuous Improvement in the Primary Classroom: Language Arts Grades K–3* (ASQ Quality Press: Milwaukee, WI, 2000).

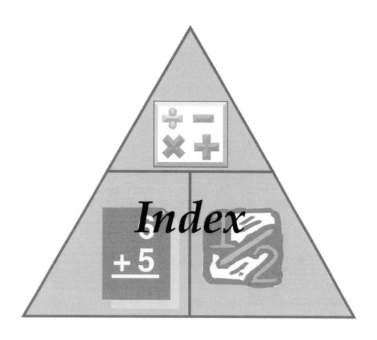

Index

Page numbers in *italics* indicate use in tables and figures.